THE FAIR SOCIETY

THE *Fair* SOCIETY

THE SCIENCE OF HUMAN NATURE AND THE PURSUIT OF SOCIAL JUSTICE

Peter Corning

The University of Chicago Press
CHICAGO AND LONDON

PETER CORNING is director of the Institute for the Study of Complex Systems and has published widely in the life sciences and social sciences.

The University of Chicago Press, Chicago 60637
The University of Chicago Press, Ltd., London

Printed in the United States of America

20 19 18 17 16 15 14 13 12 11   1 2 3 4 5

ISBN-13: 978-0-226-11627-3 (cloth)
ISBN-10: 0-226-11627-1 (cloth)

Library of Congress Cataloging-in-Publication Data

Corning, Peter, A., 1935–
The fair society : the science of human nature and the pursuit of social justice / Peter Corning.
p. cm.
Includes bibliographical references and index.
ISBN-13: 978-0-226-11627-3 (cloth : alk. paper)
ISBN-10: 0-226-11627-1 (cloth : alk. paper) 1. Fairness. 2. Social justice. 3. Basic needs. 4. Social contract. I. Title.

BJ1533.F2C67 2011
303.3′72—dc22

2010021771

∞ The paper used in this publication meets the minimum requirements of the American National Standard for Information Sciences—Permanence of Paper for Printed Library Materials, ANSI Z39.48–1992.

FOR THE SEVENTH GENERATION

Contents

Preface

> The truth has long been known and has been the bond of the wisest spirits. This old truth—reach for it.
>
> GOETHE

Long before the financial earthquake of 2008 had produced what will live in infamy as the Great Recession, deep tectonic cracks were already visible in the high temple of free market capitalism—the American economy.

Among the fissures that could be plainly seen, if anyone was looking for them: the income gap between the richest and the poorest members of our society has been growing rapidly over the past two decades. In 2007 we held the dubious distinction of having the widest income disparity in the industrialized world.[1]

More important, the housing bubble that was pumping up our economy masked an ever-spreading wasteland of poverty. Several studies over the past decade and even some obscure government statistics have shown that, well before the financial crash, close to one-quarter of our population were struggling to meet their basic needs.[2] Now the situation is much worse. One indicator is the staggering fact that close to 50 million Americans experienced "food deprivation" (hunger) at various times in 2009.[3]

Meanwhile, top executives at some of America's major corporations were engaged in what some critics have characterized as legalized looting.[4] Even the nominal owners—the shareholders—were unable to put a stop to it. Likewise, at several large, unregulated shadow banks, financial entrepreneurs were paid multimillion-dollar bonuses for selling so-called derivative securities that even they did not really understand. Trillions of dollars were sucked out of the global economy and into bundles of mortgage-backed securities, which were promoted with risk-assessment computer models that negligently excluded the historical data relating to the housing bust in the 1930s, during the Great Depression. This "inconvenient truth" was deemed not relevant in an era of ever-increasing house prices! And all this was blessed by the supposed "watchdogs" at the securities rating agencies.[5]

To make matters worse, much of this torrent of money was being invested in a ticking financial time-bomb that finally exploded—a frenzied and corrupt mortgage market that had abandoned the traditional, commonsense requirement that home buyers should have a documented credit history and the wherewithal to pay their mortgage. This was free market capitalism run amok, and the result was economic disaster.

Beyond the immediate crisis, there were (and still are) many other deep cracks in the temple walls. Consider our health care system. We spend more than twice as much per capita as any other industrialized nation and have done the poorest job of providing care for all of our citizens.[6] Close to 50 million people (one-sixth of our population) were uninsured in 2009, and at least as many more had grossly inadequate health insurance coverage.[7] Medical bills were the most frequent cause of bankruptcy in this country.

Medicare for our elderly was also in deep financial trouble, and our national health statistics have continued to erode. Most developed countries are getting much better health results at far less cost, judging by such sensitive health indicators as infant mortality and life expectancy.[8] Only our private health insurance industry is happy with this dysfunctional system, and the much-debated reform legislation that was enacted in Washington in 2010 will only drizzle on their parade. It was an important—even historic—first step in repairing the damage, but there is much more to be done.

The list could go on and on. For instance, wartime tax cuts for the wealthy (resulting in the lowest overall tax burden of any industrialized nation) have predictably led to an escalating national debt and such neglected problems as a seriously decaying national infrastructure of roads, bridges, tunnels, water systems, sewers, and so on. The deferred maintenance backlog is estimated at more than $2 trillion.[9]

We also have a public education system that is woefully underfunded and a monument to mediocrity (it ranks well below most other industrialized nations in comparative student tests), along with a public transportation system that is a national embarrassment compared with those of other developed countries and a tax system that is manifestly unfair and so complex it makes Rube Goldberg look like an engineering genius.

In short, free market capitalism has not lived up to its billing. The "land of opportunity" has become a rigged game that is deeply unfair and favors the few. Witness the taxpayer bailouts for bankers followed within months by a shameful return to bonuses as usual—"doing God's work" as Lloyd Blankfein, the CEO of Goldman Sachs, notoriously claimed.[10] (In fact, it is reported that three-quarters of Goldman Sachs's profits come from its own in-house investment portfolio.)

Witness too how the insurance industry has thwarted more serious health insurance reform and how the banking industry has resisted needed financial reforms with a lobbying campaign estimated at $150 million. The CEO of Mellon Bank (another beneficiary of taxpayer largesse) assures us that "capitalism works. Darwinism works."[11] Well, it certainly works for bank presidents who get help from the government. As the saying goes, the mega-banks have privatized their profits and socialized their losses.

All of this and more indicates clearly that it is time to rethink the "social contract," the implicit bargain among the members of any stable society that undergirds the legitimacy of the economic and political system and ensures that there will be voluntary compliance (for the most part) with its norms and laws. What the distinguished American political observer Walter Lippmann many years ago termed our "public philosophy" is at a major fork in the road, and so is our economic and political system.[12]

Communism and state-sponsored socialism were tried out in the Soviet Union and Communist China during the twentieth century and were ultimately discredited as failures. Now free market capitalism has gone over a cliff. Where it will ultimately land—and in what condition—remains to be seen. Wall Street may have recovered, but much of the rest of the country remains mired in a deep recession. And this is happening at the very time when new discoveries in the sciences have shown that the underlying assumptions that have animated capitalist doctrines and policies are in fact simplistic caricatures. The ayatollahs of economics have misled us—and themselves.

For a generation or more, mainstream (neoclassical) economists have been telling us that capitalist economies are driven by the rational pursuit of self-interest and the "invisible hand" of self-organized, self-correcting free markets—a model that goes back to the early economist Adam Smith in *The Wealth of Nations*.[13] However, it is now abundantly clear that the invisible hand can morph into a sleight of hand, as Adam Smith himself warned in an earlier work, *The Theory of Moral Sentiments*.[14]

Moreover, the new discipline of behavioral economics, which strives to determine how people actually behave in the marketplace, has discovered that the standard economic model of a rational, calculating economic man is deeply flawed. Among other things, behavioral scientists are finding that reciprocity, a sense of fairness, and even some degree of altruism are bedrock human values that also shape our economic and social behavior (as we shall see). Indeed, some of what we do routinely—like aiding others in need—could be considered highly irrational from a conventional economist's perspective.

Meanwhile, a number of other scientific disciplines have converged on the conclusion that our moral sensibilities represent an important, evolved component of human nature (discussed in chapter 4). Our moral values are based not solely on customs, or social norms, but on a complex interplay between nature and nurture.

This conclusion has been reinforced by the finding among anthropologists that a sense of fairness and concern for social justice are cultural universals. However, these moral impulses have been swamped and suppressed by the dynamics (and the incentives) in our ego-driven capitalist system. The highly touted benefits of free enterprise have been inequitably distributed, and the inevitable result is a broad pattern of social injustice. As the former labor secretary and now professor of public policy Robert Reich concluded in his book *Supercapitalism,* we need to create "new rules of the game . . . rules that reflect our values."[15]

Capitalist societies, taking their inspiration from Adam Smith and philosopher John Locke, have promoted a social contract based on a strictly limited government that serves primarily to protect individual "life, liberty, and property." Indeed, the sanctity of private property, regardless of the distribution of wealth, has been one of the cornerstones of capitalist economic theory.

Socialism, in contrast, draws its version of the social contract from philosopher Jean-Jacques Rousseau's concepts of natural equality, common property, and the "general will" of a democratic state. The modern welfare state tries to split the difference between capitalism and socialism, but this so-called third way remains a contentious battleground, and the results—certainly in this country—have obviously been unsatisfactory. We can do much better.

In this book I will propose that we erase the blackboard (or whiteboard, or smartboard, if you prefer) and start fresh with a new social contract—new rules of the game. I call it a biosocial contract because it is grounded in a biological perspective on human nature and the human condition. An organized human society can be characterized as being, in essence, a "collective survival enterprise" that is concerned primarily with the ongoing survival and reproductive needs of its members. Whatever may be our perceptions, aspirations, or illusions (or, for that matter, whatever our station in life), our "prime directive"—to borrow a term from the TV series *Star Trek*—is to provide for our basic needs and those of the next generation. This is the fundamental challenge for all living things.

A corollary of this prime directive is that the survival problem for each of us, as well as for any organized society collectively, encompasses no fewer than fourteen distinct domains of basic needs. These needs are not optional; they are biological imperatives. They have been fully documented under the Survival Indicators project at Institute for the Study of Complex Systems and are described in several of my previous publications. I will discuss these basic needs in some detail in chapter 5.

A new biosocial contract must be grounded in this biological reality, and it must strive to achieve a fair society. Our primary social (and moral) obligation is to provide for all the basic needs of our citizens. However, a biosocial contract must include other important dimensions of fairness as well. There are, in fact, three distinct fairness precepts—equality, equity (merit), and reciprocity—that must be combined into a package and balanced in order to create a society that is relatively fair and just to everyone in terms of both the benefits and the costs. Indeed, the fairness of our social, economic, and political relationships is vital to sustaining a voluntary and harmonious social contract. I will have much more to say about these three fairness precepts later on.

In relation to these precepts, neither capitalism nor socialism can be considered fair, even in theory. Each, in different ways, is unfair in its basic aims and values. Accordingly, I will propose that we consign them both to the museum of antiquated ideologies and replace them with what I refer to as the "Fair Society." The Fair Society represents a new economic and political vision that combines what I call "biosocialism" (a basic needs guarantee that fulfills our prime directive) with what has been termed "stakeholder capitalism" as well as appropriate social obligations (or reciprocities). The Fair Society is a vision that, I contend, approximates the ideal of a just society that can be traced back to Plato's great dialogue, *The Republic*. It's an "old truth" that is now also informed by modern science. To echo a signature election campaign slogan, I believe this is the change we need.

1 * *Life Is Unfair*

Life is unfair.

JOHN F. KENNEDY

Lilly Ledbetter worked for the Goodyear Tire and Rubber Company at its Gadsden, Alabama, plant for nineteen years, most of that time as an area supervisor. This was a major achievement because, as a woman in that job, Lilly was a pathbreaker. Over the years, she was able to hold her own in a mostly male (and macho) environment and even to excel, winning a "top performance" award in 1996.[1]

But Lilly's stellar job performance was not reflected in her compensation. Although she started out at a salary level comparable to that of the other area supervisors, as time went on she fell further and further behind. On several occasions she was given "poor" performance evaluations by outspokenly prejudiced superiors and was denied a raise. By the end of 1997 Lilly was making between 15 and 40 percent less than any of her male peers. Her salary level was $3,727 a month, while the salaries of the other fifteen area supervisors ranged from $4,286 to $5,236.

When Lilly finally discovered the salary discrepancy, she filed a "questionnaire" with the federal Equal Employment Opportunity Commission (EEOC) under the Civil Rights Act of 1964. Among other things, Title VII of the act prohibits wage discrimination in the workplace. Lilly did not pursue the matter at that time, but after taking early retirement in late 1998, she filed a lawsuit against Goodyear. In the trial that followed, she produced witnesses and other evidence of wage discrimination, and the jury found in her favor. She was awarded back pay, damages, and court costs.

However, the company appealed the decision, and eventually the case made its way to the United States Supreme Court, where in May 2007 the justices ruled five to four in favor of Goodyear. Writing for the majority, Justice Samuel Alito argued that the case was "untimely" because the law requires a person to file a discrimination claim with the EEOC within 180 days after a "discrete act" has occurred. Since the discriminatory pay decisions against Lilly had taken place years earlier, the Court ruled that she could not "restart the clock." Of course, it could be argued that each paycheck was discriminatory, but the Court claimed that this reasoning could not be used to "breathe life" back into the case. In other words, any discriminatory act that goes unchallenged for more than 180 days is grandfathered, regardless of its long-term consequences.

Justice Ruth Bader Ginsburg, in a stinging dissent joined by three other justices, charged that the majority was, in effect, undermining the "remedial purpose" of the Civil Rights Act and that their "parsimonious" reading of the act was cavalier about the realities of the workplace. Discrimination in matters like promotions or firings is often immediately apparent to the victim, but pay discrimination is much more subtle. It often involves repeated small acts that can have a large cumulative effect over time. Moreover, the vast majority of United States firms have explicit rules against allowing workers access to information about the compensation of their coworkers. So pay discrimination is very difficult to detect, much less to document, and employees are almost always under pressure to avoid "making waves."

The most important point in Justice Ginsburg's dissenting opinion, though, was her challenge to how the majority defined what constitutes a discriminatory act. Instead of confining it to pay decisions made years earlier, she argued, the Court should have viewed each subsequent paycheck as an unlawful act. This would have conformed more closely to the intent of the Civil Rights Act, she asserted, and thus would have been more fair-minded.

When the Court's ruling was announced, it evoked a firestorm of protest. It was headlined in the news media, denounced on various editorial pages, and quickly became a hot topic on more than half a million Internet Web sites and blogs. Congress also moved quickly on the matter, and within a month Congressman George Miller of California introduced the "Lilly Ledbetter Fair Pay Act of 2007." In late summer, the bill passed the House of Representatives by a vote of 225–197. However, it stalled in the Senate, where it failed to obtain the 60 votes needed to bring the measure to a final vote. (The procedural tally was mostly along party lines.)

One of the senators who voted for the bill was Barack Obama, who later featured the issue in his presidential campaign and invited Lilly to come to one of his rallies. After the 2008 election, the bill was approved in the new Congress by overwhelming majorities, and it was the first major piece of legislation that President Obama signed into law. (Lilly was present at the signing ceremony.)

Such acts of discrimination—of manifest unfairness—are, unfortunately, commonplace in our society and in many others, as we know very well. Sadly enough, injustice is an ancient and enduring theme in human history. So the Lilly Ledbetter case is not so unusual. But it does illustrate many things about the nature (and nurture) of fairness and unfairness.

First, there are many kinds of unfairness. Some go under the heading of "equity," or what Aristotle called "proportionate equality." Given Lilly Ledbetter's job performance over the years, it was inequitable in strictly quantitative, monetary terms that she did not receive comparable pay. But in many other cases, the issue of fairness involves conflicts between two or more parties over how to divide up a pie, or how to allocate shares. Many divorce cases are of this kind. These are preeminently situations where compromises may be required. The challenge is to find a middle ground where voluntary consent is possible and, ideally, where all parties feel fairly treated. As President Obama has put it, "My heroes are people who struggled not only with right versus wrong, but right versus right."[2]

Affirmative action in college admissions is an example of this dilemma, and we still have not found a satisfactory resolution to the issue. In the not so distant past, many American colleges routinely (but quietly) discriminated against racial minorities, as well as women and Jews. Even today, minority students, especially African Americans, often carry the twin burden of past discrimination and current economic and educational disadvantages. Yet white applicants, who are also competing for scarce college admission slots against ever-lengthening odds (the average acceptance rate for colleges with restricted admissions is about one out of six), feel discriminated against if the process is not based primarily on merit and achievement.

A similar problem exists with respect to "legacy" admissions—the offspring of older alumni. This is a time-honored tradition for many private colleges, where admissions officers are often given some "discretion." So affirmative action is not at heart a racial issue but an issue of fairness. (A recent Supreme Court case involving promotions for New Haven, Connecticut, firefighters is another example of such a fairness dilemma.)

A second lesson to be drawn from the Ledbetter case is that power very often trumps fairness. Given the economic interests at stake here and the political alignment of the conservative justices and the Republicans in the Senate, the initial outcomes were not surprising. The Court's interpretation of the Civil Rights Act can be seen as an obvious rationalization designed to support a desired objective, namely, to favor large business corporations and to constrict the scope and effectiveness of the act. But this doesn't change the fact that many of us viewed what happened to Lilly Ledbetter as profoundly unfair. The American Civil Liberties Union characterized the decision as giving corporate America a "get out of jail free" card.[3]

The Ledbetter case also illustrates how much our attitudes toward fairness are influenced by cultural norms and practices and by what other people around us, or our society as a whole, define as fair and unfair. One hundred years ago in this country, women were routinely shunted into certain categories of low-paying, low-status jobs (like telephone operators), and many people considered this perfectly legitimate. As is often the case with such blatantly discriminatory practices, a cluster of demeaning social prejudices about women were used to support and justify these actions.

Many cynics over the centuries, dating back at least to the Sophist school in ancient Greece, claim that there are no objective standards of justice. It's purely "subjective"—a matter of convention and politics, or a cloak for self-interest. The "golden rule" on Wall Street over the years has been "Whoever has the gold makes the rules." But the Ledbetter case highlights the fact that existing cultural patterns are not the whole story. The enactment of the original Civil Rights Act was itself a testament that our sense of fairness goes deeper than whatever our society currently defines as fair. The reality is that our standards of fairness can change over time in response to facts on the ground and changing perceptions and attitudes. (How and why these changes come about is a subject we will explore later on.)

Thus slavery was once a widespread, seldom questioned, and elaborately justified practice in many "civilized" societies. For more than two thousand years, from ancient Athens in its heyday to mid-nineteenth-century America and even today in a few corrupt backwater societies, slavery was a profitable international business wrapped in legal protections and supported by cultural prejudices with deep roots in human nature and human psychology, as we shall see. Slaves were often treated as subhuman or alien, like enemies in a time of war. Indeed, many slaves in ancient societies were war captives or their descendants.

It is one of the great and frequently noted paradoxes about human morality, and especially our sense of fairness, that we are very prone to draw a sharp line between "us" and "them"—those who are perceived to be inside (or outside) our group, our party, our religion, our race, our organization, or our nation—and to apply our standards of fairness selectively. Psychologists call it "moral disengagement" or "emotional disengagement," while sociologists often refer to it as the in-group/out-group phenomenon. (The treatment of minority workers in this country, both in pay scales and in working conditions, is another shameful example.) So we will also have to confront this puzzling double standard in human nature in due course.

But perhaps the most striking and important aspect of the Lilly Ledbetter case, with profound implications for our image of human nature, was the groundswell of public resentment and anger on her behalf. Many millions of Americans were offended by the perceived injustice in this case and came to her support. How come? Why is it that most of us, though not all (we'll also get to this issue later on), have an acute sense of fairness, not only for ourselves but vicariously for others we don't even know? It is called empathy, and it can produce feelings ranging from dismay and sadness to outrage and a strong desire for redress, for punishment, even for revenge. As Mark Twain might have put it, reports that a sense of fairness does not exist have been greatly exaggerated.

What is it in the human psyche that can arouse such feelings of empathy and a wish to see "justice" done for others? Contrary to the stereotype about our innate selfishness and greed, most of us share a desire to live in a society where fairness is the operative norm, where everybody's basic needs are met (polls show we are unconditionally altruistic on this score), where the economic benefits are distributed "equitably" (in accordance with merit), and where there is a robust sense of "reciprocity"—a rough balancing of benefits and obligations. What is often referred to as the "norm of fairness" depends on these principles, and it is the key to achieving a stable "social contract"—a venerable philosophical concept as well as a commonly used colloquial expression that I will resuscitate and redefine in due course.[4]

Needless to say, such feelings and attitudes are not unique to our own society. Fairness is a universal human preoccupation and an incessant daily concern in almost every society. (Again, there are pathological exceptions that I will discuss later on.) The fairness principle is also enshrined in the one great moral precept found in virtually every organized society and in every religious doctrine—the Golden Rule (Do unto others as you would have them do unto you). Indeed, something approximating the Golden Rule could be likened to a "golden thread" that binds together every voluntary, consensual relationship, from a harmonious family to a legitimate democratic government.

Fairness has also been a surefire theme in countless Hollywood movies over the years. Although our movies often present us with exaggerated, fun house mirror images of human nature, they also reflect and appeal to our deepest social instincts.

Our legal system is also based on the bedrock norm of fairness. The ideal of providing an impartial and equitable framework of public "law and order" has served as the very foundation for Western, democratic political systems. The principle of legal "justice" can be traced back to the ancient Greek lawgiver Solon, more than two thousand years ago, and it was elaborated in ancient times by, among others, the Greek Stoics, Plato, Aristotle, and the Roman lawyers, especially in the writings of Cicero.[5] Though still very far from being fully realized in practice, it is the ideal of justice—symbolized by the famous statue of the Roman goddess Justitia, who is blindfolded but holds a sword in one hand and a balance scale in the other—that inspires the elaborate and often cumbersome system we take for granted. We only notice when it breaks down.

Another surprising and important finding about human nature is that we prefer (most of us) to have a justice system that is fair to everyone (equal protection under law) and are offended when we see the system corrupted—when it has become an instrument for various economic or political interests. This is especially true when official, institutionalized injustices are rationalized with simplistic, one-sided, or even specious arguments. Indeed, the struggle to achieve an independent judiciary is still being waged—sometimes bitterly—in various countries, especially in some of the emerging nations.

This ongoing political struggle highlights one of the deeper challenges that every large, complex society faces. In the many small hunter-gatherer societies that have been studied extensively by anthropologists over the years and that are widely believed to reflect our remote history, reciprocity and sharing are the norm, and violators are punished, banished, or even assassinated. But social relationships in complex modern societies are also deeply influenced by the sometimes vast differences in power and wealth, and these disparities are very often used to distort the social contract and subvert the fairness principle—to state the obvious.

In other words, fairness is often undermined by various social, economic, and political interests, and a tacit social contract can degenerate into a more or less constricting harness, or even something approximating a concentration camp for those who are severely disadvantaged. (We will visit one infamous historical example in chapter 6.) In fact, the asymmetries produced by abuses of power are the underlying causes of many social conflicts—from labor strikes to bread riots, revolutions, civil wars, and indeed many of the lethal confrontations between nations. The patterns of exploitation and unfairness that can arise from an overreach of power and self-interest (a might-makes-right attitude) are therefore one of the major obstacles to a stable, equitable society—a fundamental insight that goes back at least to Plato's *Republic* and Aristotle's *Politics*.

Indeed, the deep injustices that can result from great disparities in power and wealth—when some people are "have-mores," to borrow a phrase from former president George W. Bush—has been one of the most important and enduring themes in the "tradition of discourse"—from the Greeks to Cicero, Saint Thomas Aquinas, Rousseau, Locke, Hume, Bentham, Mill,

Kant, Descartes, Spinoza, Marx, and, in our own time, John Rawls, Michael Walzer, Roger Masters, George Klosko, James Q. Wilson, Richard Joyce, and others.[6] Jean-Jacques Rousseau, observing the decadence all around him in prerevolutionary France, was perhaps the most passionate about it. He was appalled that "a handful of men be glutted with superfluities while the starving multitude lacks necessities."[7] We stand on the shoulders of these great moral theorists, and we can still learn from their insights. (I will briefly consider some of their views later on.)

However, something new and profoundly important has been happening during the past two decades, and it is changing the entire context for the debate about fairness and social justice. We have been witnessing the emergence of a full-blown "science of human nature," a diverse effort involving many disciplines, including evolutionary biology, neurobiology, behavioral genetics, human ethology, several branches of psychology, anthropology, economics, sociology, political science, and even the study of animal behavior. This broad, multidisciplinary effort is providing us with new insights and new perspectives on some ancient questions, and (I will argue) definitive resolutions to some long-standing philosophical and ideological debates.

In a nutshell, we are beginning to get a fix on the deep structure of human nature. Equally important, we are gaining new understanding about how our cultures mold and shape what biologists call the "phenotype"—the learned and socially conditioned behavior of the people we interact with every day. We now also appreciate more fully that we are products of a very long evolutionary process. Our remote ancestors and ancestral species lived in small, socially organized, and intensely interdependent groups for perhaps several million years, and both our genetic heritage and our cultural heritage reflect this cardinal fact. The implications of this important scientific enterprise are profound and merit a full-length review in a later chapter. Here I will provide only an abbreviated overview.

In general, our sense of fairness appears to be a joint product of both nature and nurture, very like the acquisition of language skills. The norm of fairness first appears at a very early age. It involves, in essence, the recognition of entitlements that apply to others as well as to oneself. Simple decision rules like equal shares, taking turns, or drawing straws work well enough with young children. But as a child develops, the content of the sense of fairness changes and deepens, as a rule, and more complex criteria are added—age, merit, need, even our social relationships and "we/they" distinctions.

Needless to say, the content of what is viewed as fair is also influenced by the values, customs, rules, and practices of a given society—what others believe is fair. And, of course, we also have a propensity to rationalize unfairness away when it suits our interests, perhaps with a twinge of guilt. Nevertheless, as a rule fairness has a strong, though imperfect, pull on our perceptions and our conduct, although there are always significant personality differences that must be accounted for, as we shall see.

To illustrate, consider the classic story *The Little Red Hen,* one of the all-time best-selling children's books. The little red hen works hard and is frugal. One day she finds some grains of wheat and decides to plant them. She asks her friends (a dog, a cat, and a pig, in one version of the story), "Who will help me plant these seeds?" Well, her friends all have more important things to do, so she plants them herself. And so it goes at each successive stage in the process—tending and weeding the garden, harvesting the wheat, threshing the grain, grinding the flour, and baking the bread. At each step the red hen asks for help, but her friends are always too busy. Yet when it finally comes time to eat the bread, her friends are more than willing to help (no surprise). They're eager to have a share. By then, of course, it's too late. (I trust you know the outcome.)

Now, a Marxian philosopher or a well-trained defense lawyer might object that the red hen should not have eaten all the bread herself, but many generations of children, unburdened by the teachings of our moral philosophers and legal scholars, seem to have gotten the point on their own. It's a matter of reciprocity and equity. It's not fair for the red hen's friends to receive a share of the bread if they were unwilling to help produce it. (Our economists and game theorists would call them "free riders.")

The hard evidence that a norm of fairness and reciprocity is a universal aspect of human nature can fairly be called robust. As I noted earlier, it's found in virtually every society, albeit with variations, and the pathological exceptions seem to prove the rule. Fairness is a day-in, day-out issue in every society. There is also a large experimental literature on this phenomenon in psychology, game theory, anthropology, and behavioral economics, among other disciplines.

Most noteworthy, perhaps, are the "ultimatum games," which have demonstrated repeatedly that people are willing to share with others in ways that do not reflect their own narrow self-interest but instead demonstrate a sense of fairness. Equally important, it appears that people are far more willing to invest in policing fairness and punishing deviants than classical economic theory predicts. Particularly significant is the work on "strong reciprocity," which highlights altruistic aiding and punishment behaviors.

There are even some rudimentary examples of a sense of fairness in other socially organized species, the most conspicuous of which are sharing behaviors and reciprocity in primates (see chapter 4). Finally, the accumulating psychological evidence of a sense of fairness has been given an evolutionary underpinning with the resurgence of "group selection theory" in evolutionary biology, most notably in the work of biologist David Sloan Wilson and his colleague Elliott Sober.[8] Our "social instincts," as Darwin called them, are very likely a part of our evolved human nature(s) and not simply a cultural overlay (a key point that I will pursue further in due course).

In short, the standard neoclassical economics model of a rational, calculating, acquisitive economic man (*Homo economicus*) is a caricature that obscures a much more complex reality. It's only part of the story, more likely to be true for some individuals in some contexts than for others. Many if not most of us (according to the research data on this subject) can more accurately be characterized as *Homo reciprocus* (after Howard Becker) or *Homo reciprocans* (after Samuel Bowles and Herbert Gintis). Our social contract is based on reciprocity, mutual aid, and even some targeted altruism, not on unbridled competition. When you stop to think about it, the vaunted "markets" that lie at the heart of every capitalist society are, at bottom, elaborate and highly structured systems of reciprocity (of exchange): competition is only a supporting actor, or sometimes a necessary evil.

The new science of fairness has many implications, but certainly one of the most important has to do with our public philosophy—our common understanding about the way our society works and what kind of society we want to live in. For the past 150 years or so, the debate on this issue has been dominated by two simplistic, one-sided doctrines—capitalism and socialism. Both are based on flawed and distorted assumptions about human nature and are deficient in their understanding about the basic purpose of an organized society. And both have been responsible for inciting some of the most colossal human tragedies of our era.

To be sure, there have been various attempts over the years to formulate a "third way"—a middle ground between capitalism and socialism that could embrace and reconcile the two competing ideologies. The third way amounts to being a matchmaker for an arranged marriage between defenders of free markets and private wealth, on the one hand, and advocates for the social and "safety net" programs of the welfare state on the other.

In light of the science of fairness, such a marriage of convenience is inadequate. To summarize the argument I will develop later on, both capitalism and socialism are partly right but also seriously wrong about human nature. Both discount the ability of culture—our norms and expectations—to shape the "rules of the game." Both are also rife with self-fulfilling doctrines. Both are insensitive to, and sometimes myopic about, the important variations in human nature and the influence of specific cultural contexts, which represent wild cards in any social contract. And both are at once fair in some ways but also deeply unfair in others.

The punch line of my argument against capitalism and socialism is this: Capitalism lavishly rewards self-interest and competitive success but is indifferent to any external standard of fairness and reciprocity, while the satisfaction of our basic, biologically grounded needs is taken for granted, or discounted, or shrouded in sophistry. Capitalism favors honesty and fair dealing in principle, but it cannot really justify such moral precepts, much less fairness and social justice.

Socialism, on the other hand, comes in many different flavors with various political agendas, but the common core is an egalitarian concern for our basic needs. A more equitable distribution of wealth is a common objective, but socialism can also be blind to the claims of merit—talent, hard work, and achievement. Capitalism is concerned about the values of freedom and the sanctity of private property, whereas socialism is concerned about freedom from want, and freedom from fear, for the members of society as a whole—to paraphrase President Franklin Roosevelt's famous "Four Freedoms" speech in 1941. (I will flesh out these abbreviated "bullet points" and give them a more nuanced treatment in a later chapter.)

One alternative to having to choose between capitalism and socialism is to start over with a revisioning of the entire paradigm. In light of an evolutionary and biological perspective on human societies and, equally important, the growing power of the new science of human nature, I propose to call it a "biosocial contract." Though the concept of a social contract is obviously borrowed from traditional political theory, a biosocial contract is a different animal. It is not a hypothetical construct or a debating device but a shorthand description of the values and practices that characterize any reasonably successful and sustainable society.

To summarize this new vision very briefly, the ground-zero premise (so to speak) of the biological sciences is that survival and reproduction constitute the basic, continuing, inescapable problem for all living organisms: life is at bottom a "survival enterprise." (Darwin characterized it as the "struggle for existence.") Furthermore, the problem of survival and reproduction is multifaceted and relentless; it is a problem that can never be permanently solved. Thus an organized, interdependent society is quintessentially a "collective survival enterprise." To borrow a term from sociobiology, it's a "superorganism."[9]

This taproot assumption about the human condition is hardly news, but we very often deny it, or downgrade it, or simply lose touch with it. The claim that a society is merely a facultative arrangement—a marketplace or perhaps a vehicle for material and moral improvement—downplays or even denies its true nature. Our fundamental collective purpose is to provide for the basic survival and reproductive needs of our people—past, present, and future. In effect, we are all parties to a biologically based contract. (In fact, humans are not unique in this respect; other socially organized species in the natural world also have an implicit biological contract.)

This biological contract, and the imperatives (and rules) associated with it, encompasses the preponderance of human activity, and human choices, worldwide. To be sure, survival per se may be the furthest thing from our conscious minds as we go about our daily lives. Nevertheless, our mundane daily routines are mostly instrumental to meeting the underlying survival challenge. They reflect the particular strategy—the package of cultural, economic, and political tools—by which each society organizes and pursues the ongoing survival enterprise.

Accordingly, we are endowed with an array of biologically grounded "preferences" (in the argot of economics) that are virtually universal, and we mostly choose to follow their dictates. Moreover, these preferences are not all created equal. This allows us to seek regularities, make "if-then" predictions, and link human nature to human behavior, including our social ethics, in comprehensible ways.

Very briefly, the first and most important generalization about human nature is that each of us is defined, in considerable measure, by an array of "basic needs" that are essential to our survival and reproductive success, and we come into the world with an orientation toward satisfying these needs. The concept of basic needs is hardly new, needless to say. Its roots go back at least to Plato and Aristotle, and it has been used in various ways over the years, ranging from a narrowly focused preoccupation with food, clothing, and shelter to psychologist Abraham Maslow's expansive claims for "self-actualization."[10] More recently, social scientists Len Doyal and Ian Gough have advanced the claim, in *A Theory of Human Need*, that participation in the life of the community is our ultimate objective, and that personal health and "autonomy" are the necessary means.[11]

However, such psychological and ethical definitions of basic needs downplay and even discount the ongoing challenge of biological survival and reproduction. More relevant is the growing body of empirical research, most notably under the sponsorship of the United Nations, the National Academy of Sciences, the World Bank, and other agencies, that gives scientific credence and considerable precision to the concept of basic needs.

In addition, the so-called Survival Indicators project at the Institute for the Study of Complex Systems (see chapter 5) has involved an effort to develop and validate measuring rods for the full range of survival requisites for an individual or a population.[12] In the current version of this framework, no fewer than fourteen "primary needs" domains have been identified and documented. These represent biological imperatives in any given society or personal situation, in conjunction with an indeterminate number of context-specific "instrumental needs." (Primary needs are irreducible and vary within well-defined parameters, but instrumental needs vary widely and are often highly localized in nature.) Needless to say, this broad formulation cuts a very wide swath through any complex economy.

Within the context of this fundamental biological enterprise, there exist in every stable society both implicit and explicit expectations about the behavior of others (customs, norms, rules, and laws) that provide a framework for the social, economic, and political relationships that sustain the ongoing survival effort. Capitalism is focused on the pursuit of our material self-interest and on fulfilling our personal preferences for goods and services. Socialism is focused on the common needs of the community as a whole—and the role of the state as an instrumentality in some formulations.

However, a biosocial contract is focused on our *relationships*—the social interactions that occur every day among individuals, families, organizations, neighborhoods, communities, and businesses and between citizens and their government(s). And it is the quality, and fairness, of these relationships that largely determines how effectively any given society is able to pursue its collective survival enterprise. In fact, any complex, interdependent economy depends on cooperation (what I prefer to call a "combination of labor" as opposed to the economists' division of labor), as well as on mutual aid and—most important—reciprocity and reciprocal obligations. Needless to say, it matters a great deal whether these social relationships are voluntary, equitable, and durable or coerced and potentially unstable. Thus fairness is an inescapable problem and an important part of the solution for every viable society.

So how do I define fairness within the framework of a biosocial contract? What are the ground rules or guiding principles? Later on, I will develop and defend the proposition that there is in fact a nested set of three distinct precepts, or directives, that must be implemented in a balanced and fair-minded way in any society. They are, first, an unqualified commitment to ensure that all the basic (survival) needs of every member are fully provided for; second (beyond that), an equitable distribution based on "merit" of whatever surplus of goods and services a society produces; and third, a reciprocal obligation that requires everyone (with some important exceptions) to contribute an equitable share to the collective survival enterprise and to support the practices, norms, and laws that undergird it.

How do we go about implementing this paradigm? In the final chapter, I will propose that we replace capitalism and socialism—those dog-eared and increasingly dysfunctional nineteenth-century ideologies—with a new ideological synthesis. It entails a combination of what has been termed "stakeholder capitalism" and a basic needs "guarantee" (what I call biosocialism), along with a strong obligation for reciprocity. Also important is what has been called "community governance," or cooperative local bootstrapping efforts.[13] I call this new formulation the "Fair Society" model. (The term is not new, but I will use it in a significantly different way. See endnote 14 for this chapter.)[14]

Such a revisioning of our public philosophy is obviously a very large, complex, and controversial undertaking. It raises many issues and invites many questions, which I will address in due course. As we proceed, some sacred cows will be gored and some vested interests will be divested. There is much work to be done.

2 * The Idea of Fairness

> *David Drumlin*: You must think this is all very unfair. What you don't know is that I agree. I wish the world was a place where fair was the bottom line . . . unfortunately, we don't live in that world.
>
> *Ellie Arroway*: Funny, I always believed that the world was what we make of it.
>
> FROM THE FILM *Contact* (1997)

What is fairness? And why do we care? Why is fairness such a timeless and universal human preoccupation?

Let's start with President John F. Kennedy's famous remark about fairness. At a press conference in March 1962, a reporter asked him about some army reservists who had complained publicly when they were called up for active duty. After defending his decision as being vital to United States foreign policy, Kennedy became a bit philosophical: "There is always inequality in life. Some men are killed in a war and some men are wounded, and some men never leave the country, and some men are stationed in San Francisco. It's very hard in the military or personal life to assure complete equality. Life is unfair."[1]

With all due respect to a martyred and revered American president, this was not a very good example of unfairness. Assuming that these soldiers either volunteered or had been drafted in a fair (random) selection process, their obligations to serve were unconditional and their assignments were based on their training and the needs of the military. There was, presumably, no personal animus or discrimination in the decision to call them up for active service—no bias was involved.

In contrast, consider the systematic discrimination against African American soldiers and loyal Japanese American (Nisei) volunteers during World War II. Blacks not only were segregated but were routinely assigned to the dirtiest and most dangerous support jobs. There were even some shameful incidents where German prisoners of war were given preference over our own black soldiers. Likewise, Nisei combat units were often singled out for the most difficult battlefield assignments and suffered very high casualties. Why were such actions unfair? Because patriotic Americans were being deliberately discriminated against for no good reason—out of racial prejudice. Their treatment was inequitable (unwarranted) compared with the treatment of others in the military.

And that's the point. If the "vicissitudes of life" are a result of decisions that were even-handed and perceived to be legitimate and fair, then we may or may not be happy about them, but we do not as a rule "take it personally." If others have treated us fairly, we generally comply and accept our good (or bad) "luck" fatalistically. To cite one example, during the current war in Iraq our soldiers have been asked to make extraordinary sacrifices, and most of them have stoically accepted their grueling, life-threatening assignments and their multiple tours in combat. (The griping among the army reservists in President Kennedy's day soon abated.) Likewise, if we are the victims of impersonal forces—a chance occurrence or circumstances beyond anyone's control (say an earthquake, or a flood, or a wildfire where some houses burn to the ground and others are spared)—we do not ordinarily think about it in terms of fairness.[2]

This illustrates once again one of the cardinal points about fairness. Fairness is all about how we are treated in our *relationships* with others. It is always a personal matter—an aspect (or a property) of the associations and interactions we have with other people, or with various organizations and institutions. Fairness means taking into account and balancing the needs, interests, and rights of all parties. The question in every situation is whether or not there was fair dealing and equity both in the process—how everyone was treated—and in the distribution of benefits and costs for everyone concerned. What were the consequences?

If, for example, I am motivated by the current high price of gold to purchase some gold prospecting equipment and I then go out looking for gold without hitting "pay dirt," the fault is my own. I would most likely call it either bad luck or a bad decision. But if a seller of gold prospecting equipment persuades me to buy it because he claims to know where gold can be found and says it's a sure thing when in fact he knows it is not, then I was intentionally misled by someone who benefited while causing me "harm" (a key term that we will come back to again and again). Sadly enough, such deceptive practices are common in our capitalist society, and the more serious cases are rightly treated as civil or criminal offenses. They violate the norm (and the psychology) of fairness—not to mention the rule of law.

In other words, the particulars of each case, and people's intentions toward one another, are crucial factors in how we make our fairness assessments. If somebody is acting in good faith toward us and has no personal interest or stake in our actions, we are likely to excuse him or her even if we were unintentionally harmed—say, when somebody gives us bad driving directions that lead us astray. We will assume that he or she was at least trying to be helpful.

On the other hand, if a person had something to gain from misleading us, we are likely to feel betrayed. This is why the currently popular "conflict of interest" and "full disclosure" policies have become so important in recent years. Transparency provides a tool for ensuring fair dealing—or allowing us to opt out. It creates a cultural bias toward fairness. On the other hand, the many deliberately deceptive and misleading practices that American corporations and banks undertake in the name of profits are often harmful and deeply unfair. One token example is the way food packagers have been putting ever smaller portions into ever larger boxes with ever increasing air space. More serious cases involve the knowing sale of defective or dangerous products.

So the essential first principle in every social relationship, and in every (stable) human society, must be honesty and fair play. Deception, lying, cheating, stealing, or any other action that deliberately causes harm to another person is (among other things) an example of unfair conduct. Ultimately it undermines the implicit social contract based on mutual trust and reciprocity that we all ultimately depend on in our social relationships. (I'll talk much more about the social contract later on.)

It is also important to differentiate between objective fairness criteria, including perhaps various quantitative measures, and our often foggy perceptions, attitudes, and emotions about fairness. To use the terminology of social psychology, our experience of fairness has both a "cognitive" element and an "affective" element. Our feelings about fairness—what we are predisposed to believe is fair—represent an important factor. But these predispositions are, in turn, heavily influenced by the people around us and by our culture. As a species with a deep history as social animals, we are highly sensitive to our social environments and, by and large, are keen to get along with the people around us. The psychologists tell us that we are therefore very prone to accept and accommodate to what others expect of us, especially if they are authority figures.

A particularly disturbing illustration of this is the famous (and recently replicated) experiment conducted by psychologist Stanley Milgram in the 1960s. Experimenters in lab coats were routinely able to persuade their subjects to administer what they thought were painful electrical shocks to other subjects behind a glass wall, supposedly as a learning technique. (The victims were actors, and no actual shocks were inflicted.)[3]

Another example of our susceptibility to social influences is a behavioral effect that Milgram also identified—what is sometimes referred to as the "aggressive triad." Aggressive behaviors are promoted when one person has the power to make the decision without having to execute it while a subordinate can perform the act without having to be responsible for it. It has been said that there would be fewer wars if the "decider" (as President George W. Bush liked to call himself) also had to do the fighting. No wonder, then, that fairness is such an elusive and context-specific concept, and no wonder so many skeptics discount it. Yet the new science of fairness has shown that it is very real and very important, as we shall see.

Our dictionaries define fairness in more or less straightforward ways. We are told that fairness refers to "equitable, honest, impartial dealings" (*Oxford English Dictionary*), "freedom from self-interest, prejudice, or favoritism" (Merriam-Webster), "conformity with rules or standards" (Word Net), and "free from bias, dishonesty, or injustice" (Dictionary.com).

The problem, of course, is deciding what is equitable, free from bias, impartial, and so forth, in any given situation. It can be a very tough call. Even such moral absolutes as "Thou shalt not kill" or "Thou shalt not steal" have their ambiguities. Thus many people seem to think it is perfectly all right to kill your enemies in wartime. And fictional heroes like Robin Hood and Zorro have had an enduring appeal for their dashing, selfless acts: stealing from the rich and giving it to the poor. It is much the same with fairness.

There is a delightful 1998 Irish movie called *Waking Ned Devine* that captures some of these ambiguities. In the movie, the elderly Ned dies of a heart attack when he watches a televised drawing and learns he has won the national lottery. Two of his close friends, discovering his body with the winning ticket still clutched in his hand, conspire to keep his death a secret (he has no family) while one of them poses as the dead man to collect the lottery winnings. Technically it's fraud, of course. But as in so many fictional (and real-life) stories, the question of fairness, and our attitude toward it, puts a different spin on the story.

In the movie, Ned's two friends begin to have second thoughts about what they are doing—and so does the audience—when they learn that the jackpot is worth £6 million. After some soul searching, they decide the money should be shared equally among all fifty-two members of their small (poor) Irish village, where there are many unmet needs. That's what Ned (who did win, after all) would have wanted to do with the money, the two friends conclude. In a curious way, their new plan to share the wealth with their needy community, rather than hoarding it for themselves, makes the scheme seem less reprehensible. It will do much good for others.

The problem is that all fifty-two members of the community must be willing to go along with the plan, because a representative from the lottery will be coming to the village to verify the dead man's identity. And unfortunately there is one holdout—a mean-spirited miser (nicknamed "the witch") who rides around in a motorized wheelchair, though she is perfectly able to walk, and who is notorious for insulting and offending people and for behaving like Scrooge. She withholds her approval from the scheme and demands £1 million as her share. Otherwise she threatens to report the plot to the authorities and collect a reward amounting to 10 percent of the winnings.

However, the townsfolk refuse to give in to her "blackmail." She is in the very act of calling the lottery headquarters from a public pay phone on the edge of a cliff overlooking the sea when the village priest, driving back from a trip, swerves to avoid an oncoming car and knocks the phone booth off the cliff. (Was this the hand of God?)

Now, the point of this story is that the villagers (and most of the audience, I surmise) view this outcome as a happy ending, even though it entailed both fraud and a person's accidental death. How can this be? Our legal tradition says we should condemn and bemoan these acts, while neoclassical economic theory (not to mention neo-Darwinian "selfish-gene" theory) says we should admire the miser for using her economic leverage to get as much as she could for herself. This sort of thing happens in free markets every day. The answer is that something else is going on here that modifies our attitude toward our legal and economic (and evolutionary) theories. It involves our innate sense of fairness—our sense of justice.

In short, fairness is not some sort of cosmic absolute or a moral Jell-O mold. It is always defined by the particular context—the people who are the participants and their *relationships* and *interactions*. There are very often two (or more) sides to any fairness issue, and sometimes a disinterested third party is needed—a friend, a parent, a counselor, a mediator, an arbitrator, a judge, or a jury—to break the impasse and adjudicate between various fairness claims. As the great American judge and legal scholar Learned Hand put it many years ago, "Justice is the tolerable accommodation of the conflicting interests of society."

Of course, human cultures have also evolved a variety of formulas, or social tools, to help people decide what is fair in the welter of everyday fairness situations. Providing "equal shares" for everyone is one of the oldest and most reliable practices. It is a useful solution whenever there is a fixed pie to be divided up. Everyone is treated as equal, and the outcome is impartial.

Another common fairness practice is queuing (or first come, first served), as well as using various (impartial) games of chance such as drawing straws, dealing cards, and the like. To cite one example, in a recent local election in the state of New Mexico, where there was a tie vote for the town mayor, the deadlock was broken (in accordance with state law) by flipping a coin, and the winner of the toss was declared the victor. (In Arizona, by contrast, they use a deck of cards.) The importance of such everyday fairness facilitators is underscored when somebody violates them. There are few things more infuriating to the driver of a car who dutifully changes lanes when directed to merge than having other cars come from behind, go to the front of the line, and then cut in.

Some of the hardest fairness calls in real-life situations are cases where our traditional rules of thumb don't work—when equal shares is not really appropriate and when "equity" (or proportionate shares) is really the fairest way. Consider the timeless phrase that every parent has heard so many times, "It's not fair." Is it fair to give a five-year-old child and a teenager the same allowance?

On the other hand, if there is a birthday party for your fifteen-year-old, and her five-year-old brother is invited so he will not feel left out, is it really appropriate to give him an equal-sized portion of the cake? But just try to reason with him about it. A wise parent will not even attempt to be "equitable." One solution is to give everyone an equal share and save the five-year-old's leftovers for later. Or, if the cake is big enough (if it's not a "scarce resource," as our economists would say), another option is to let the guests choose as much or as little as they like. In effect, a free choice, whenever possible, disarms our propensity to draw comparisons and to become envious or resentful.

Many societies also modify strict fairness rules with various handicapping procedures. Thus, in the United States young children and the elderly are often allowed to go to the head of a line, or to pay reduced fees for various activities, and people who are physically handicapped have been provided with various special accommodations, thanks to the Americans with Disabilities Act. Most of us accept such favoritism without giving it much thought, because it seems intuitively to be fair. Yet there are inconsistencies. Even millionaires get senior discounts, which the rest of us must subsidize. But then, the number of millionaire seniors is minuscule compared with the number of low-income seniors who also benefit. (Golf handicaps, which are designed to offset widely varying individual abilities, are another commonplace example.)

Are such handicapping principles fair? The answer is that they are fair if we believe they are fair. Our sense of fairness may ultimately be labile and "subjective," but it's no less real. As we shall see, it's ubiquitous, and it's fundamentally important. It affects just about every corner of our vast worldwide web of human relationships. Indeed, there are many different arenas and many different levels where fairness issues are played out each day. Let's talk briefly about some of these.

Fairness (and unfairness) starts with our most intimate personal relationships—our friends, loved ones, families, and communities. We also meet fairness issues every time we purchase or sell some item, every time we go to work or participate in some activity with others, and every time we have to deal with some formal institution. Social scientist Jon Elster and his colleagues have developed a multidisciplinary field called "local justice," which studies in depth how fairness issues are dealt with in such cultural hot spots as kindergarten and college admissions, military drafts, the demand for kidney transplants, adoption, child custody awards, and the like. The complexities associated with many of these commonplace fairness problems can be daunting.[4]

Fairness is also the focus of many contentious economic and political issues. There is, for example, the "fair trade" movement, which has been attempting to implement more equitable international trade policies. Such organizations as Oxfam, Amnesty International, Catholic Relief Services, and Caritas International have been helping disadvantaged producers in emerging nations to achieve better returns for their businesses, as well as providing better working conditions and improving environmental standards. Much effort is also being made to obtain more equitable terms in the formal trade agreements between nations, where differences in bargaining power can seriously disadvantage a smaller, weaker partner.

In economics, there is a large body of theory and practice on "fair pricing." When no price yet exists for an item, various methods have been developed for establishing what will be viewed as a reasonable price under various circumstances. There are also the methods used in the accounting field to determine "fair value" on an item when there are no relevant market prices to use for comparison. The concept of a "fair profit" is important in cases where there is no competitive bidding and "cost-plus" contracts are used.

Then there is the concept of a "just price," an ethical concern tracing back to the ancient Greeks and Saint Thomas Aquinas in the Middle Ages that has also been associated with the injunctions against usury in the Christian Bible and the Muslim Quran. (Originally, both religions considered it sinful to charge any interest at all on a loan, but modern doctrine is more lenient. These days usury commonly means charging interest that exceeds legal limits or is considered "excessive." Some of the so-called payday lenders these days, charging up to 400 percent on an annual basis, are usurious by any standard.)[5]

Though now defunct, the "fairness doctrine" in broadcasting, from the 1950s to the 1980s, represented yet another attempt at fair dealing. It involved a requirement that every broadcaster, as a condition for using the public airwaves without cost, was obligated to provide free airtime for controversial issues and to present them in a manner that was "honest, equitable, and balanced." Some would like to see this rule reinstated, but the current political climate seems to favor creating greater broadcasting diversity rather than reimposing any compulsory measures. There is also the seldom used "equal-time rule," which requires any station that gives free airtime to a candidate for elective office during prime-time hours (except for regular newscasts and news events) to grant an equal amount of time to any opponent who requests it.

Needless to say, our democratic political system also involves a thicket of fairness issues, from voting restrictions (there have been many battlegrounds over the years, including age, sex, poll taxes, literacy tests, etc.) to the apportionment of legislative districts, lobbying activities, legislative procedures, campaign fund-raising, and much more. Indeed, the core idea of majority rule, one of the oldest and most sacred of democratic principles, also reflects the principle of fairness. How so? Because everyone has an equal share in making a decision.

Of course, in modern legislative bodies there is usually a power structure—party organizations, leaders, committee chairmen, seniority rules, and the like—that make some votes more equal than others. And in the United States Senate, where each state, regardless of size, is entitled to two senators, the principle of one man, one vote is egregiously violated. In 2008 Wyoming had a population of 533,000 and California had some 36.7 million. A California senator represents roughly sixty-nine times as many people as a Wyoming senator, but she still gets only one vote in the Senate. Then there is the filibuster rule in the Senate, which allows a minority of forty-one senators to block the majority vote if they choose to use it (which seems to be the rule these days).

The principle of proportional representation (which is practiced in many countries) involves an especially significant but little-appreciated example of a fairness issue in politics. In many election systems, the operative rule is winner take all; whichever candidate or party obtains a majority controls the outcome, and the minority is completely disenfranchised. This is how we (mostly) do things in our political system, and how the Republican Party apportions convention delegate votes in its primaries. In contrast, the Democratic Party, in the interest of fairness, allocates its convention delegates in accordance with the breakdown of the primary vote in each state, along with using various formulas to apportion caucus votes. This is one reason the Democratic primary season in 2008 was so extended, and so exhausting. (On the other hand, the Democrats' policy of awarding "superdelegate" votes to influential party members like governors, congressmen, and former presidents evoked controversy in 2008 because some perceived it as undemocratic and unfair.)

Fairness is also an aspect of virtually every legislative, regulatory, and judicial decision in our society. These actions almost always differentially affect different interests, sometimes in large numbers. Call it mega-fairness.

A prime example is President Harry Truman's aptly named "Fair Deal" program after World War II.[6] In the 1948 election campaign, Truman called for the enactment of a broad array of social welfare programs designed to benefit the poor and the middle class. However, his proposals were met with fierce conservative opposition (the Republicans then controlled the Congress), and most of them finally fell victim to the Korean War in 1950. Truman's most notable success was the public housing program of 1949. Nevertheless, he laid out the social welfare agenda for generations to come. Some of Truman's other proposals were finally enacted under President Lyndon Johnson's "Great Society" program in 1964. Still others were accomplished, or augmented, by Presidents Jimmy Carter and Bill Clinton, and even by some Republicans. But one of the most important of Truman's proposals, universal health insurance, remains to be achieved sixty years later.

War is an arena in which fairness issues are not given priority, to put it mildly. The old saying "All's fair in love and war" is certainly true for the most part in modern warfare (although the record concerning romance seems to be more mixed). Nevertheless, various constraints and "rules of the game" have been exercised in warfare from time to time, and certain kinds of weapons have been banned by international agreements. In the 1899 Hague Convention, for instance, poison gas, chemical weapons, and hollow-point bullets were prohibited. (The ban on dropping explosives or projectiles from balloons became moot after airplanes were introduced during World War I, and poison gas was used for a time in that war despite the agreement but was finally abandoned voluntarily by all sides.)

The treatment of prisoners of war has also become subject to explicit rules under a series of Geneva Conventions, along with a variety of "laws of war" and actions that constitute "war crimes." Yet the recent transgressions by our own country illustrate how fragile these constraints remain. It seems that wars often induce the combatants to abandon the normal rules of social intercourse. One of Winston Churchill's more famous sayings illustrates the point: "In wartime, the truth is so precious that she should always be attended by a bodyguard of lies."

Fairness is obviously also an important issue in sports, where (as legend has it) the word "fairness" was first coined by a British sportsman in the nineteenth century. In fact, the very expression "to be a good sport" implies acting fairly and winning or losing graciously. The proverbial "level playing field" (a favored expression in the business world) is a metaphor borrowed from sports. And the rules of every organized sport—which are sometimes excruciatingly detailed—are designed for the most part to avoid giving anyone an "unfair" advantage. This is why lead weights are used to offset the weight differences among jockeys, why tennis players must regularly switch courts during professional tennis matches (to balance any sun and wind advantages), why there are coin tosses before football games, why the "home field advantage" is offset by games played "on the road," why there are different weight classes in boxing and wrestling, and why there is a much-abused ban on performance-enhancing drugs in sports, among other things.

Fairness is also a principle that is fundamental to a "free press." Ideally, press freedom allows journalists to be impartial, and it is a time-honored tradition in this country that reporters should try not to take sides in their news articles, in contrast to editorial writers. We all know that our news media fall short of this ideal, sometimes egregiously so. (Fox News, which claims to be fair and impartial, is a blatant example.) But many journalists—sometimes at great personal risk—have fought the good fight over the years, and few of us appreciate the debt we owe them for their struggle to report the truth even-handedly. (A recent Hollywood movie paid tribute to the courageous stand that Edward R. Murrow, one of our great TV journalists, took in the 1950s against the anti-Communist bullying and persecution by Senator Joseph McCarthy.)

Fairness is also deeply embedded in our legal system, needless to say. As we shall see, the concepts of fairness and justice were virtually synonymous back when the latter term (*dikaisyne*) was first used by the ancient Greeks (our "justice" comes from the Latin words *ius* and *iustus*), though today "fairness" and "justice" have distinct though overlapping meanings. Over the past 2,500 years we have evolved a highly elaborate formal system of justice that includes a complex set of institutions, processes, and procedures along with a vast repository of legal precedents, many of which have established new norms and new expectations with regard to fairness.

Consider just a few of our foundational legal principles in this country: equality before the law, equal protection of the laws, due process of law, habeas corpus, our constitutional prohibition against cruel and unusual punishment, and much more. Our legal system is festooned with rules and practices intended to ensure fairness and impartiality in everything from the procedures used for picking juries to the ethical requirements imposed on judges, the right of an accused person to defense counsel, and the controversial "Miranda ruling" by the Supreme Court, which requires the police to inform suspects of their constitutional rights when they are arrested.

Consider also the ancient Anglo-Saxon legal principle that an accused person must be presumed innocent until proved guilty. The burden is on the accuser (very often it's the state these days) to prove guilt "beyond a reasonable doubt." Though Americans take these legal safeguards for granted (or become disillusioned when they are sometimes betrayed or corrupted), they do not even exist in many other legal systems, where the law is little more than an instrument for maintaining political control.

But what is generally referred to as "procedural fairness," in all its many forms, pales in importance compared with "substantive fairness"—the huge, complex, and ever-changing battleground where many of the outcomes in life are decided and distributed. We can break down substantive fairness into three broad categories. These will come to play a central role later on in the formulation of a biosocial contract and my proposals for a Fair Society.

Equality, or "equal shares." This is the most fundamental principle of substantive fairness and, by and large, the easiest to administer. (We met it at the teenager's birthday party.) Equal shares also accords with a strong human desire to be treated equally. Yet even here there are ambiguities. Equality before the law, for example, is not applied to children in our legal system. Rather, we have developed a largely separate juvenile justice system with procedures and penalties that are sensitive to the immaturity and inexperience of young people (though in certain states there have recently been measures to treat some youthful offenders as adults). Equality is also important in any cooperative team effort, where everyone contributes, though perhaps in different ways, to the achievement of a collective goal. But the principle of equality makes its greatest claim on our resources with respect to our basic biological survival needs. There is no substitute for equality where these survival imperatives are concerned, for anything less results in doing a person more or less serious "harm" (that word again). This is a concrete floor where everyone is approximately equal and where everyone is equally deprived if any basic need is not satisfied.

Equity is a more complex and challenging concept, but it is no less fundamental. There are a great many situations in human societies where equality is not generally perceived as being fair—where there is an inequality in the equation. This idea marches under various banners—"justice," "just deserts," giving all persons their "due," or (again) Aristotle's "proportionate equality." However, I prefer to use "merit," which implies that a reward, or punishment, is earned by the recipient and is a result of his or her actions and circumstances. Of course, this provokes a question that can never be answered once and for all because it is so dependent on the particulars of each case. How do you determine what a person merits? In capitalist economic theory, for instance, merit is closely associated with talent, initiative, private investment, risk taking, hard work, and, of course, achievement. Failure is seldom generously rewarded—except in some recent cases where CEOs have received opulent salaries and bonuses just before their companies went bankrupt or were bailed out by the taxpayers. Capitalist theory is pointedly indifferent to whether such actions are fair. For the most part, the apologists for capitalism point to the success stories and focus on efficiency, progress, and economic growth. It's a cliché going back to Adam Smith himself that a competitive marketplace—with winners and losers and greed galore—ultimately benefits society as a whole. (The important social science research in what is called "equity theory" will be discussed later on. See also endnote 7.)[7]

The "dark side" of merit has to do with letting the punishment fit the crime. Whenever people fail to do their share, or are free riders on the efforts of others, or engage in cheating, or in some way fail to live up to a contractual agreement, they are in effect exploitative and cause "harm" to others. You do not have to be a fan of capitalism to see that it is unfair to give equal rewards to a hard worker and a slacker. (Remember the little red hen?) This is what merit pay is all about, to take one example. And this is why, both in human societies and in many animal societies (as we shall see), punishments are a common way of dealing with "defectors"—as the game theorists call them. This is also why humans have such highly developed sensibilities about cheating and free riding, and why we are quite willing to punish offenders, albeit with some important exceptions. But there are also those among us who exemplify the classical model of *Homo economicus*—people who ruthlessly pursue their self-interest, often at others' expense. Capitalist economists and libertarians pay lip service to the idea that freedom is not a license to harm others, but they mostly trust in the "invisible hand" of the marketplace and its workings to keep everyone within bounds. As we have recently seen, this is a self-serving myth. Unbridled self-interest, when coupled with disparities in economic or political power, can trample fairness sometimes literally to death.

Reciprocity is the third but no less important fairness domain. When a disciple of Confucius asked him for a single word that would describe the basic principle of social life, he is reputed to have answered "reciprocity." Reciprocity is most often associated with fair *exchanges* of goods and services, or "tit for tat," but it also means that you must pay back a kindness or a favor. Otherwise, you are tacitly exploiting others and benefiting unfairly from their labors. As the great Roman legal scholar Cicero put it, "There is no duty more indispensable than returning a kindness."[8]

Not only is the "norm of reciprocity," in sociologist Alvin Gouldner's term, a universal ethical principle in human societies, but rudimentary examples can be observed in primate societies as well (see chapter 4). Humans keep score of the favors they do one another, though the material value is often less important than the value of these exchanges in building and maintaining personal friendships and mutual trust, and in signaling a willingness to cooperate and assist one another. Sometimes a simple expression of genuine appreciation can be reciprocity enough. Reciprocity is, in fact, a form of psychological glue that binds together our friendships, our families, and our communities. (The extensive research on reciprocity in recent years will also be discussed in chapter 4.)

Threaded through all these categories of fairness are variations on the fundamental principle of the Golden Rule. At a Parliament of the World's Religions in 1993, a proclamation signed by 143 leaders from the world's religious and spiritual communities declared the Golden Rule to be a universal norm. Of course, this only ratified what was already widely recognized. Although the basic idea has been expressed in many different ways over the centuries, the Golden Rule stands out as a moral precept that has global acceptance.[9] Here are just a few selected examples:

> *Confucius:* "Do not to others what you do not want done to yourself—this is what the word [reciprocity] means." (*Analects,* 15.23)
>
> *Judaism:* "You shall not take vengeance or bear a grudge against your countrymen. Love your fellow as yourself." (Leviticus 19:18)[10]
>
> *Hinduism:* "Do not unto others what ye do not wish done to yourself; and wish for others too what ye desire and long for, for yourself." (Vedic scriptures, *Mahabharata*)
>
> *Christianity:* "And as you would that men should do to you, do you also to them likewise." (Luke 6:31, quoting Jesus)
>
> *Zoroastrianism:* "What I hold good for myself, I should for all." (Zoroastrian writings, Gatha, 43.1)
>
> *Islam:* "The noblest religion is this, that thou should like for others what thou like for thyself; and what thou feel as painful for thyself, hold as painful for all others too." (Hadis, Sayings of Muhammad)
>
> *Jainism:* "Just as pain is not agreeable to you, it is so with others. Knowing this principle of equality, treat others with respect and compassion." (*Suman Suttam,* verse 150)

Countless renderings of the same basic idea can be found in the writings of many of the world's great thinkers—Thales, Plato, Aristotle, Epictetus, Saint Thomas, Locke, Mill, and Spinoza, to mention just a few. Still other theorists seem to have endorsed the same basic principle without acknowledging it as such. Let's look at two important examples.

One is Immanuel Kant, considered by some admirers to be the greatest moral philosopher of all time. Kant is perhaps best known for what he viewed as a self-evident foundation for morality based on unaided reason alone, the "categorical imperative." The categorical imperative states, quite simply, that you should "act only according to the maxim whereby you can at the same time will that it should become a universal law." Kant claimed that "there is, accordingly, no need of science and philosophy to know what one has to do in order to be honest and good."[11]

Kant's specification for such a categorical imperative was that it must be right for everyone and that everyone must be treated as an end and not as a means to an end. The Golden Rule, Kant claimed, does not satisfy these conditions because some actions taken under this principle might not be considered universally right. As an example, according to Kant, if under the Golden Rule you were a judge and did not yourself want to go to prison, how could you sentence anyone else to do so?

Of course, if you want to live in a society governed by rules and laws that are applied to everyone impartially—"universally"—and where transgressors are fairly and consistently punished, and if it is your sworn public duty to uphold and enforce these laws, you might conclude that it is a categorical imperative as well as a legal obligation to punish anyone who commits a crime, including yourself. This is not inconsistent with the categorical imperative, or the Golden Rule.

Kant's objection to the Golden Rule is especially suspect because the categorical imperative sounds a lot like a paraphrase, or perhaps a close cousin, of the same fundamental idea. In effect, it says that you should act toward others in ways that you would want everyone else to act toward others, yourself included (presumably). Calling it a universal law does not materially improve on the basic concept. The well-known economist and game theorist Ken Binmore (who claims to have read almost everything Kant wrote), comes to the same conclusion. "It eventually dawned on me that I was reading the work of an emperor who was clothed in nothing more than the obscurity of his prose. . . . His categorical imperative is simply a grandiloquent rendering of the folk wisdom [contained in the Golden Rule]." Kant's justifications are "conjured from the air."[12]

The other philosopher who seems unwittingly to have borne witness to the Golden Rule is John Rawls. In his celebrated 1971 book *A Theory of Justice*—a complex, provocative, and much-debated work that broke new ground in political theory—Rawls mounted an effort to justify the claim that justice should be defined in terms of fairness.[13] (As we shall see, this aligned him with Plato and Aristotle to a greater extent than he perhaps realized.) Rawls did not propose to do away with economic inequalities. Instead, he posited two broad principles: equality in the enjoyment of personal freedom (consonant with the freedom of others), and a set of economic arrangements that would allow for equal opportunity coupled with ways to permit the poor, or the "least advantaged," to benefit proportionately more when the rich get richer. Moreover, any economic inequalities should be of benefit to everyone—a provocative idea that raises the question: How would this work, precisely? Maybe Rawls was thinking about the now badly tarnished idea of a "trickle down" of new wealth.

Rawls's method for justifying and supporting these principles was at once ingenious and frustrating. He asked us to assume that we are in a hypothetical "state of nature" before the existence of society—an "original position" in which we are all behind a "veil of ignorance" about our own station in life. Rawls called his approach an elaboration on Kant's categorical imperative, but this is not so. In what amounted to an appeal to enlightened self-interest rather than dispassionate reason, Rawls argued that his principles are those we would rationally choose for organizing our society if we were uncertain what our own circumstances might end up being. So Rawls's prescription really amounted to the Golden Rule in deep disguise. Once again, Binmore concurs. He says that Rawls's original position is "just an elaboration of the ubiquitous Golden Rule." Binmore also called Rawls's invocation of Kant "window dressing." (For more discussion of Rawls, see endnote 14.)[14]

Of course, there are also a great many cynics and self-proclaimed "realists" who simply reject the Golden Rule altogether. Thus, the crusty twentieth-century playwright George Bernard Shaw declared that "the golden rule is that there is no golden rule." Paraphrasing one of the fundamental assumptions of capitalist theory, Shaw asserted that other people's tastes "may not be the same [as ours]."[15] Various philosophers, including Friedrich Nietzsche, Bertrand Russell, Robert Nozick, and others have made similar arguments. How can we know what other people may want?

The commonsense response to these objections is that we don't need to know a priori. Wouldn't you want to have a Good Samaritan take account of your specific needs and wants in helping you, just as you would want to be sensitive to his or her needs if your situations were reversed? But more important, the Golden Rule is not just about gift giving. It embraces all the fairness precepts I talked about above—honesty and fair dealing, equality, equity, and reciprocity. It covers the totality of our relationships with one another as they play out in different contexts. So there is a great deal of common ground at the core of the Golden Rule.

Given the universality of the Golden Rule, it is surprising, to say the least, that very few of its proponents over the millennia have ever taken the trouble to explain in any detail why it should be treated as a general moral principle. Why is it so unequivocally and unquestionably true? The reason, I believe, is because it is consistent with our own intuitive (and I will argue innate) sense of fairness, plus something more that we will explore further in due course. We are, most of us, also endowed with an evolved sense of altruism toward others regarding what could be called "no-fault needs" (to borrow an insurance industry term), needs that arise from the sort of vicissitudes of life I talked about earlier. In other words, the Golden Rule resonates with our innate sensibilities; it plucks a string in the human psyche.

To summarize, then, fairness is about the quality and content of our relationships with one another. It refers to both the processes and the outcomes of our social interactions. It is deeply affected by our expectations for ourselves and others, as well as by what we perceive to be others' intentions toward us. Though our sense of fairness exerts a strong centrifugal (altruistic) pull on our behavior, we are also subject to the centripetal (inward-looking) forces of self-interest. When this self-absorption and self-regard are coupled with disparities in economic and political power, it can have a corrosive effect on fairness, as we shall see. And when unfairness becomes endemic, it can cause great harm in terms of our collective needs or can even produce a breakdown in the social contract.

How do we know when fairness has been achieved? Often enough it's an unattainable ideal, but it can be approximated when all parties to a fairness issue are fully informed and voluntarily agree that the result is reasonably fair under the circumstances. Anything less is in varying degrees coercive, exploitative, harmful, and in extreme cases lethal.

So here are some examples of everyday fairness issues. How would you respond to these questions?

- Is it fair to keep a convicted prisoner incarcerated if DNA evidence later proves he or she is innocent?[16]
- Is it fair for our schools to promote students who fail along with the students who work hard and get good grades?
- Is it fair for a hardware store to raise the price of snow shovels in anticipation of a spring snowstorm when supplies of shovels are low (to use a textbook example)?
- Is it fair for somebody who has access to confidential and privileged information about a publicly held corporation that will affect the price of its stock when it is made public to take advantage of it and engage in "insider trading"?
- Is it fair for a qualified auto mechanic who made a mistake in repairing your car to charge you for a return visit to correct his mistake?
- Is it fair for some of our millionaires and billionaires (and corporations) to set up offshore tax havens so they can avoid paying taxes?
- Is it fair that capital gains from stock investments should be taxed at a lower rate than ordinary income from work?
- Is it fair for the banks that were bailed out by the taxpayers to return to paying out billions of dollars in bonuses to their employees while the

many millions who have been thrown out of work as a result of their malfeasance are struggling to meet their basic needs?

If you answered no to all of these questions, it's apparent that you have a finely tuned sense of fairness. I trust you will not be disappointed with what's to come. If you answered yes to any of them, I hope I can persuade you to change your mind.

Unfortunately, some of us are seriously deficient in the fairness domain. A sense of fairness is a personality trait that is evidently not equally distributed, which is one reason there are so many fairness issues surrounding us. In the chapters ahead we will explore further where our sense of fairness comes from, how it has played out historically, and why some of us seem to be moral eunuchs. We will see that fairness has been an age-old struggle, one that has been regularly discounted in theory but is ever present in fact. It is a moral principle that, it seems, each new generation has to redefine within the context of its own time and place, and yet there are also some unchanging universals.

3 * *A Brief History of (Un)Fairness*

> Any State, however small, is in fact divided into two—one the State of the poor, the other that of the rich—and these are [forever] at war with one another.
>
> PLATO, *The Republic*

The Golden Rule is hardly the rule in modern societies. If we just look around us, Plato's gloomy description of the human condition seems to ring true, at least as a metaphor. Fairness is certainly not a term that comes to mind when we consider the breadth and depth of poverty and suffering in our country and around the globe and contrast it with the enclaves of vast wealth that seem to be insulated from what most of us are experiencing in life.

In fact, there are many radical individualists who reject the very idea of fairness; to them it's an alien concept. For instance, a neo-Darwinian evolutionary theorist, such as Richard Dawkins in his influential book *The Selfish Gene,* would argue that the operative principle in the natural world—the basic law of nature—is that every organism must act in its own self-interest; it must maximize its chances of reproductive success. As Dawkins put it, "We are survival machines—robot vehicles blindly programmed to preserve the selfish molecules known as genes. . . . I shall argue that a predominant quality to be expected in a successful gene is ruthless selfishness. . . . we are born selfish."[1]

In a similar fashion, the neoclassical economists, inspired by Adam Smith's vision in *The Wealth of Nations,* have asserted that self-interest is the guiding principle in economic life. A rational person should seek at all times to increase his or her personal wealth. Likewise, modern libertarian philosophers like Robert Nozick, Ayn Rand, and others proclaim that all of us should have the freedom to pursue our self-interest, so long as it does not "harm" others. Nozick speaks of "natural justice" and asserts: "Individuals have rights, and there are things no person or group [or state] may do to them (without violating their rights)."[2]

Then there is Ayn Rand, the high priestess of libertarianism who glorified selfishness and idealized the role of the creative genius in her best-selling novels and in her so-called Objectivist philosophy. In the words of her defiantly heroic character, architect Howard Roark in *The Fountainhead* (during his courtroom trial for blowing up a building project because the owner made modifications to his design):

> Nothing is given to man on earth. . . . He can survive . . . by the independent work of his own mind or as a parasite fed by the minds of others. . . . The basic need of the creator is independence. . . . To the creator, all relations with men are secondary. . . . the creator is the man who stands alone. . . . All that which proceeds from man's independent ego is good. All that which proceeds from man's dependence upon men is evil. . . . The egotist in the absolute sense is not the man who sacrifices for others . . . he does not exist for any other man—and asks no other man to exist for him. . . . The first right on earth is the right of the ego. Man's first duty is to himself. . . . His moral law is to do what he wishes, provided his wish does not depend *primarily* upon other men. . . . The only good which men can do to one another and the only statement of their proper relationship is—hands off! . . . . Civilization is a progress toward a society of privacy. . . . Civilization is the process of setting man free from men.[3]

Critics of Ayn Rand over the years have dismissed her strident egoism and her warped romantic visions, yet her work has continued to be a potent influence in conservative circles. Her novels have sold more than 25 million copies over the past sixty years, with continuing sales of more than a hundred thousand each year. A 1991 survey of readers jointly sponsored by the Library of Congress and the Book-of-the-Month Club found that her 1957 magnum opus, *Atlas Shrugged,* ranked second only to the Bible as the most influential book of our time. Many corporate CEOs view *Atlas Shrugged* as an inspiration, according to an informal survey reported in *USA Today* in 2002.[4] And as recently as January 2009, an op-ed piece in the *Wall Street Journal* by conservative writer Stephen Moore claimed that Ayn Rand's vision in *Atlas Shrugged* of a catastrophic economic collapse presciently foretold our current economic crisis. What's more, she correctly diagnosed the underlying cause—an oppressive and rapacious government![5] (For a critique of Rand, see endnote 3.)

As for Ayn Rand's Objectivist philosophy, there are two endowed chairs in her name at American universities; there is an Ayn Rand Institute that continues to promote her work; and there is an academic journal devoted to explicating and debating her theories. But perhaps most significant, many prominent Americans acknowledge having been influenced by her writings. They include Alan Greenspan (yes, that Alan Greenspan), conservative Supreme Court Justice Clarence Thomas, former SEC chairman Chris Cox (the same agency that somehow failed to detect Bernard Madoff's Ponzi scheme), former congressman Bob Barr (who captured about 5 percent of the total vote in the 2008 election as the Libertarian Party candidate), Congressman Ron Paul (who ran a quixotic presidential primary campaign in 2008 as a "different" Republican), as well as Rush Limbaugh (yes, that Rush Limbaugh), Billie Jean King, Angelina Jolie, and many others. (A 2009 book about her, *Goddess of the Market* by Jennifer Burns, only adds to the mystique.)

In addition to these Ayn Rand camp followers, there is a much larger conservative universe consisting of (among others) the *Wall Street Journal*, the *National Review*, and *Forbes* and their readers, Fox News viewers, conservative think tanks, subscribers to conservative journals, conservative bloggers and conservative (neoclassical) economists, many of whom believe, unashamedly, that the rich are rich because they deserve to be and the poor are poor mostly through their own failings. In any case, many conservatives would claim that poverty is none of their business, and they resent having to pay taxes to alleviate economic hardship. As conservative economist Paul Rubin expressed it, "In today's world . . . people mostly become wealthy by being productive and creating benefits for others, and, therefore, desires to punish or penalize the wealthy [say by increasing their taxes] are misguided."[6]

So how do we reconcile the principle of fairness, and the Golden Rule in particular, with such an unabashedly individualistic and self-centered ideology? More important, how does it fit with Plato's blunt assessment about a "class war" between the rich and the poor, which seems to be as true today as it was in Plato's time? To gain some leverage on this fundamental issue, we need to begin with some perspective on our evolutionary heritage and the emergence of large, complex modern societies.

The "state of nature" is a theoretical device that many philosophers, dating back to the ancient Greek Sophists, have used to anchor their assumptions and speculations about human nature and the basic purpose of human societies. I identified a modern version of this in chapter 2 with John Rawls's "original position."

However, we now know quite a bit about what the state of nature was *really* like for our remote ancestors of five to seven million years ago, and it did not resemble at all the imaginative fictions of our philosophers. The origin and evolution of our species puts a very different spin on these basic philosophical issues (and on Ayn Rand's diatribes, I might add). Although there is still much uncertainty about many of the details, and though we are mindful that new fossil finds may continue to modify both the time line and the shape of our family tree (or more likely our family bush), I believe that enough of the environmental context and the basic survival challenges are now known for us to construct a plausible scenario.

Let's start with a parable, based on three baboon field studies.[7] It is early morning on the East African savanna, and a troop of sleeping baboons begins to stir. The troop has been sequestered overnight at a nesting site—the jagged face of an ancient rift—that is well protected from various carnivore enemies: lions, leopards, hyenas, and especially a marauding pack of wild dogs. But now the troop—comprising about forty males, females, juveniles, and infants—is ready to set off together on their daily food quest. The group will need to cover several miles before returning before nightfall to the safety of its sleeping site, and the members must stick together. It's dangerous for baboons to travel alone in open country.

The troop leader is a large, confident alpha male who has earned his job the old-fashioned way. He bears a number of battle scars from the formidable canines of his rivals, not to mention an occasional violent conflict with a rival troop, mostly over watering holes, feeding patches, and sleeping sites. As the troop forms a wide, irregular semicircle in a clearing below their nesting site, the leader squats in front of them on his hind limbs—erect, calm, and attentive. One by one the other baboons, using body language and tentative movements, in effect cast their votes for various alternative foraging routes. Today there seems to be a consensus in favor of a westerly route that had recently yielded an abundance of berries, fruits, nuts, roots, tender leaves and grasses, some easily captured small game, and several watering holes. The alpha male seems hesitant, perhaps remembering a recent run-in with a pride of lions. But finally he rises and begins to move. Then the leader and the rest of the troop—including several females with clinging infants and rambunctious juveniles—sets off toward the west in search of food.

Later that day, as part of the troop is negotiating a long, shallow ravine between two open areas with low shrubbery and lush grassland, one of the females, with her infant clutching tightly to her back, becomes distracted by some succulent rhizomes (rootlike plant stems that are rich in nutrients). She stops to feed, unaware that a cheetah has been stalking the group from the ridge above. Now the cheetah moves ahead of her, getting into position to cut her off from the rest of the group. At the last minute the infant—already able to recognize a dangerous predator—stiffens and utters a cry. The mother swirls around and shrieks an alarm call. The cheetah hesitates, and in the next instant half a dozen baboon males with canines bared come rushing to the mother's aid. The cheetah, finding the tables suddenly turned, hastily retreats while the baboon mother and infant quickly rejoin the group.

This episode is just part of a fairly typical day in the life of these large quadrupedal monkeys, with their distinctive doglike snouts. Life on the ground in East Africa, especially away from the safe haven of tall trees and sleeping cliffs, is both rich in food resources for omnivores like baboons and fraught with life-and-death challenges. But these "smart monkeys," as baboon expert Shirley Strum calls them, are resourceful, adaptable, and formidable.[8] Next to humans, baboons are the most successfully adapted of all the terrestrial primates. They are ubiquitous throughout Africa and the Near East. Chimpanzees, by contrast, have mostly stayed close to the safety of the trees, at least in recent times, and have relied on a narrower range of resources. They are relicts on the verge of extinction.

Although humans and chimpanzees are more closely related to each other biologically than either one is to the other primates, in some ways humankind also resembles the more tightly organized and aggressive baboons. Our baboon parable could also be a tale about our remote protohominid ancestors of several million years ago, though there were also some differences. *Homo sapiens* evolved in a rich but variable and unforgiving environment, which we have ultimately come to dominate. Perhaps the best all-around way to describe it is with Darwin's expression the "struggle for existence." Sometimes the living was easy, but many times it was not.[9]

In gradually abandoning the relative safety of the trees, probably because the forests were shrinking, and adopting a radically new survival strategy as wide-ranging ground foragers, our earliest direct ancestors had some severe disadvantages. They were very small (less than three feet tall); they were relatively slow moving, because they were still not very proficient at bipedalism; and they did not have the natural weapons that baboons can deploy—their large, menacing canine teeth.

Nevertheless, over time our ancestors were able to develop a variant of the baboon strategy. They formed small, cooperative social groups built around coalitions of closely related males that foraged together and, as necessary, fiercely defended themselves and the group against their many predators and competitors, including, from time to time, other protohominid groups. This cooperative social strategy was indispensable. There were no fewer than ten large carnivore species roaming East Africa in those days, compared with just two today, and pack-hunting species like *Palhyaena* would have been particularly dangerous. As we saw in our baboon parable, becoming separated from the group, even briefly, could be fatal. Most likely these "Miocene midgets," as anthropologist Milford Wolpoff calls them, were also aided by crude wood and bone implements that served both as digging sticks and as defensive weapons.[10] Even chimpanzees are proficient as toolmakers and tool users.

This fundamental shift in our ancestors' basic survival strategy, which of course did not happen all at once, has been called neo-Lamarckian, after an important naturalist who was a predecessor of Darwin. Though Lamarck guessed wrong about the mechanism of transmission between generations (Darwin also made some wrong guesses), one of his major insights about the natural world was that behavioral innovations often precede and precipitate biological evolution. The distinguished modern biologist Ernst Mayr referred to such behavioral innovations as evolutionary "pacemakers."

In other words, over the course of several million years, the human species in effect invented itself through an entrepreneurial process involving gradual cultural inventions that changed our relationship with the environment—and with one another. And these changes, in turn, led to the natural selection of supportive anatomical and psychological traits. As biologist Jonathan Kingdon summed it up in the title of his insightful book, we are the self-made man.[11]

The outcome of this multimillion-year evolutionary process, which occurred roughly in three distinct stages or phases, was a species that pursued its basic survival enterprise in closely cooperating and intensely interdependent groups with a high degree of sharing, reciprocity, and mutual aid—especially where food and defense were concerned. It was a strategy that also favored the evolution of such pro-social attributes as empathy, the willingness to aid the others in your group, and a concern for the harmony and well-being of the group as a whole. Call it patriotism. Thus a human society can accurately be called a "collective survival enterprise." Its basic purpose, and the foundation of our implicit "social contract," is close cooperation in providing for the survival and reproductive needs of the group as a whole—fulfilling our *prime directive.*

To be sure, our evolving hominid ancestors still displayed much individual competition and conflict, especially among the males over dominance and mating privileges. Nor does this scenario deny the existence of inequalities in intelligence, abilities, and experience. Some individuals always excelled, but these differences were harnessed to the production of mutual benefits and what economists call "public goods," or the public interest. And the inevitable social tensions were muted and constrained for the most part by a combination of our evolved internal psychological restraints and "policing" by coalitions of other group members. In a worst case, a serious offender might be physically punished, ostracized, or even killed. In other words, the egoistic side of human nature was caged and socialized, not eliminated.

Supporting evidence for this evolutionary scenario can be found in the many field studies by ethnographers over the years among the small hunter-gatherer societies that have survived in remote areas down to the present day. Anthropologist cum primatologist Christopher Boehm, in an important synthesis drawing on forty-eight of these field studies, found that there was a consistent pattern of close cooperation, sharing, and reciprocity in these societies and relatively few differences in material possessions. He also identified what he labeled a "reverse dominance hierarchy."[12] As a rule, leaders and other aggressive individuals were effectively constrained in various ways by other members of the community, and when anyone attempted to become too assertive, he (or she) was actively resisted.

In other words, the tendency to compete for power and personal advantage—and occasionally to cheat—did not disappear from the human psyche, but successful small societies of the kind that characterized our ancestors for millions of years developed very similar cultural means for controlling these potentially destructive influences. (This conclusion was recently supported by a major multidisciplinary synthesis of the empirical evidence for the last glacial period, from 74,000 to 11,500 years ago.)[13] An appropriate metaphor for this egalitarian cultural pattern is what sociologists Alexandra Maryanski and Jonathan Turner called "the social cage."[14]

Senior economist John Gowdy, in his edited collection of classic papers on hunter-gatherers, *Limited Wants, Unlimited Means,* points out that "the more we learn about hunter-gatherer cultures, the more we realize that the value system of modern market capitalism does not reflect 'human nature.' Assumptions about human behavior that members of market economies believe to be universal truths—that humans are naturally competitive and acquisitive and that social stratification is natural—do not apply to many hunting and gathering peoples. . . . The view of human nature embedded in Western economic theory is an anomaly in human history."[15]

A legendary example, thanks to the pioneering research of anthropologist Julian Steward in the 1930s and several follow-up studies since then, is the Great Basin Shoshone.[16] The Great Basin, in the American Southwest, is a dry, harsh environment, and the Native Americans who inhabited this area until very recently survived mainly by foraging in very small family groups. The bulk of their diet was plant foods—nuts, seeds, tubers, roots, berries, and the like. Opportunistic hunting of occasional large game animals (deer, mountain sheep, antelope, elk, bison) was also part of their foraging repertoire, along with some fishing and the capture of small animals. But meat probably provided less than 20 percent of their total calories.

However, there were two important exceptions to this small family foraging pattern. At times, especially during the winter months, several Shoshone families would gather in larger camps near a common resource like a water supply or a pine nut grove, which they would share. Here they would also share information, learn skills from one another, and find mates. And nobody dominated the proceedings.

But the more significant exception was the occasional gatherings in groups numbering seventy-five or more, whenever there were special opportunities for one of their famous Shoshone rabbit drives (and sometimes antelope drives). These events involved highly coordinated efforts using huge nets, rather like tennis nets only hundreds of feet long, which were deployed to encircle and trap large concentrations of prey. However, the successful use of this primitive technology depended on a division of labor between trained net holders and skilled "beaters" under the leadership of an experienced and trusted "rabbit boss." As Steward noted, enough animals were captured by these team efforts to feed everyone for many days, whereas individual efforts would have been largely ineffective. Yet even the rabbit bosses received only their share.

So early human societies were most likely characterized by (relative) internal peace and an egalitarian sharing of the means of subsistence. And an evolved sense of fairness served to lubricate the collective survival enterprise. Leadership was also important, but it played a circumscribed social role. It was not a source of personal privilege, and disruptive dominance behaviors were resisted. In other words, the egoism that Ayn Rand extolled was effectively "caged" for the common good.

But if close social bonds and socially imposed constraints were the norm within each group, quite the opposite was true of the interactions between groups, which were sometimes in direct competition. Here distrust, suspicion, antagonism, even violent conflicts occasionally occurred, and hostility to outsiders and other groups became the norm. (Again, we can find parallels in other primates, especially chimpanzees.) If fairness refers especially to achieving outcomes and distributions of goods that are arrived at voluntarily and by consensus, then coercion against neighboring groups is unfair by definition.

It is increasingly evident that, as Darwin himself supposed in *The Descent of Man*, differential selection among competing hominid groups also came to play an important role in our evolution as a species, and the psychological substrate we have inherited is in part a result of these two contradictory life-and-death influences.[17] The accumulating evidence—which I will discuss in the next chapter—suggests that both ethnocentrism (loyalty to one's own group) and xenophobia (hostility toward other groups) are evolved psychological propensities in humankind with a strong biological foundation and deep evolutionary roots. On the one hand, we may be willing to sacrifice our very lives for our metaphorical "brothers" and "sisters," yet we are equally ready to kill our enemies without qualms, without pity, and without remorse. As one of the officers portrayed in the World War II TV docudrama *Band of Brothers* told one of his soldiers, "All wars depend upon it."

Contrary to the myth of the noble savage, or the idea that human evolution was largely a peaceful process, there is reason to believe that violent conflicts between hominid groups also have ancient roots and played an increasingly important role, especially in the later stages of human evolution. It seems likely that warfare, or at least the use of coercion, was a key factor in the spread of modern *Homo sapiens* out of Africa and around the globe, beginning about fifty thousand years ago.[18]

I hasten to add that we are not talking here about wars of conquest or imperialism in the modern sense. The humans of the terminal Pleistocene were not necessarily more "warlike" in temperament, or seeking dominion for its own sake. It's more likely that the process was driven by a pressing need for resources to support a growing human population in an ever-changing environment. The last glacial period began about seventy-five thousand years ago, intensified about thirty-three thousand years ago, and peaked about twenty thousand years ago. It could be called the resource acquisition model of warfare.

The reasoning behind this hypothesis is as follows: Our late Pleistocene ancestors were not so different anatomically from other contemporary hominid species, like archaic *Homo sapiens*. But they did have superior language skills and a cultural package that gave them a decided advantage—what could be called an "imbalance of power." (Social commentator and author Malcolm Gladwell would characterize it as a "tipping point.")[19] These advantages very likely included a larger number of warriors, more effective organizational and communications abilities, and superior technology—including an advanced and more effective hand ax and, possibly, thrown spears.

The most important reason for suspecting that a pattern of warfare was deeply involved in the human diaspora is that, for the most part, these expanding human populations were not migrating into uninhabited territories. Like the fifteenth-century Europeans who discovered that their New World was already populated with "natives," the prehistoric groups that migrated out of Africa some fifty thousand years ago found that other hominids had preceded them by hundreds of thousands, if not millions, of years and had already occupied the most bountiful locations. Peaceful trade relations can arise when there is the possibility of mutually beneficial exchanges between groups. But an attempt to seize another group's territory and resources is a zero-sum game. The invaders were competitors who threatened the livelihood of the established residents. They would not have been warmly welcomed, to put it mildly.

A chilling illustration of this dynamic was cited by Jared Diamond in his Pulitzer Prize–winning book *Guns, Germs, and Steel*.[20] It involved the total destruction of the Moriori hunter-gatherer society on the Chatham Islands in the Pacific in 1835 at the hands of nine hundred well-armed Maori agriculturalists from nearby New Zealand. The Maori first learned of the peaceful Moriori from a transient Australian seal hunter. Excited by the report that the Moriori had no weapons, the Maori immediately organized a seaborne invasion. When the unsuspecting Moriori did not resist, the Maori raiding party slaughtered them with impunity. Ironically, the two groups traced their ancestry back to a common Polynesian origin, but they had long since lost contact.

If warfare played a major part in our final emergence as a species, it is also deeply implicated in the evolution of larger, more complex civilizations. Indeed, many theorists believe warfare was the prime mover. The well-known biologist Richard Alexander, for instance, advanced a Darwinian "inclusive fitness" explanation. He characterized war as reproductive competition by other means. "At some point in our history the actual function of human groups—their significance for their individual members—was protection from the predatory effects of other human groups. . . . I am suggesting that all other adaptations associated with group living, such as cooperation in agriculture, fishing or industry, are secondary."[21]

Another proponent of the warfare school of cultural evolution is anthropologist Robert Carneiro. His theory is more subtle (it relies on a functional argument rather than a presumed instinctual urge), but it too is monolithic. "Force, and not enlightened self-interest, is the mechanism by which political evolution has led, step by step, from autonomous villages to states." Although state-level political systems were invented independently several times, warfare was in every case the prime mover, Carneiro asserted. He blamed it ultimately on population pressures—what he called "environmental circumscription."[22]

It is unquestionably true that organized warfare has been a major influence in the evolution of complex societies. An analysis of this issue some years ago, which examined twenty-one cases of state development ranging in time from 3000 BC to the nineteenth century AD, found that coercion was a factor in every case and that outright conquest was involved in about half of them.[23]

However, the various deterministic, single-factor causal theories about humankind, especially ones that rely on warfare, are insufficient. If warfare involves grave and possibly fatal risks to the combatants, we need to probe more deeply into why wars occur. In fact, there is a vast body of research on this subject, spanning several academic disciplines, and this data trove supports at least one unambiguous conclusion. Warfare is itself a complex phenomenon with many potential causes and many different consequences. Wars cannot simply be treated as the expression of an instinctual urge or uncontrollable population pressures. There are too many problems with any such prime-mover theories. (See endnote 24.)[24]

The emergence of more complex human societies, beginning perhaps three hundred thousand years ago, was in fact a multifaceted process, the key elements of which can perhaps be distilled into four major categories—settlements, surpluses, specialization, and size. In turn, these four broad categories can be broken down into an array of more specific factors. Many years ago, the archaeologist and science writer John Pfeiffer noted that the rise of complex societies seemed to be closely associated with what he referred to as evolutionary "hot spots."[25] These were locations that possessed a package of needed resources—concentrations of large game animals (or other protein sources like fish), plenty of edible plant materials, ample supplies of freshwater, an abundance of firewood, and—as the agricultural revolution gained momentum—such things as fertile soil, irrigation water, a favorable growing climate, well-developed trading routes, defensible terrain, and of course technology.

Jared Diamond, in *Guns, Germs, and Steel,* details many of the cultural developments that were also part of the package, including genetically altered cereal grains and pulses, domesticated sheep, goats, and draft animals, irrigation systems, tools for plowing and for harvesting, threshing, and grinding grains, cooking implements and food storage vessels, record-keeping techniques, defensive walls, and more.

All these necessary elements combined to create a "package" of preconditions for a sharp break with our egalitarian, hunter-gatherer heritage. The combination of permanent settlements, abundant agricultural surpluses, increasing specialization of different tasks and roles, and rapid population growth altered the basic structure of human societies and, equally important, the interests and relationships of their members. Ironically, the very factors that contributed to our economic progress as a species also created the preconditions for exploitation and systemic injustice.

As anthropologist Brian Fagan points out, these cultural changes may well have begun even during the later Stone Age, when complex hunter-gatherer societies, such as the affluent Native Americans of the Pacific Northwest, became sedentary and began to display more elaborate social divisions and marked disparities of wealth.[26] Permanent settlements had the advantage of eliminating the time and labor costs required for periodic migrations, but they also created opportunities for accumulating personal goods. (Of course, they also created attractive targets for covetous neighbors.) Reliable food surpluses also provided the wherewithal for developing political and military professionals and crafts specialists, and the work of the few (farmers) could be used to support the many.

In other words, it was the very surpluses produced by emerging ancient states that lay at the root of the problem of economic exploitation. The so-called Big men and chiefs in early agricultural societies began to develop hierarchies of political control and tax systems (and systems of conscripted corvée labor for public works) backed by force. The once tight-knit, closely cooperating small communities expanded into much larger, more impersonal population centers organized around markets, with many specialized roles and many potential conflicts of interest.

When the farmers who produced the food and other raw materials were able to receive reciprocal benefits in return, including protection from external enemies, internal law and order, and the means to obtain desirable products from other specialists, this new division of labor was likely to have been perceived as equitable. Many early civilizations, particularly when they were still young, seemed to enjoy a degree of shared affluence and internal peace. But all too often this changed over time as rulers and various political and economic elites became increasingly exploitative. As the distinguished anthropologist Bruce Trigger points out in his magisterial synthesis *Understanding Early Civilizations,* "A defining feature of all early civilizations was the institutionalized appropriation by a small ruling group of most of the wealth produced by the lower classes."[27]

There is, however, one important exception to this dismal history, namely the underappreciated case of the so-called Harappan civilization—although it might be more appropriate to call it the Indus civilization. This unique though short-lived society spanned a huge territory that included the majestic Indus River and five major tributaries. It extended well into what is today Pakistan and Afghanistan. In fact, the scale of the Harappan/Indus civilization dwarfs all the other ancient societies. It encompassed about 1.2 million square kilometers, more than twelve times the size of the Nile Valley and Mesopotamia combined.[28]

What sets the Harappan/Indus civilization apart is that (apparently) there never was a centralized state with a ruling class, a military establishment, or a pronounced concentration of wealth. Instead, it represented a sprawling commonwealth with a widely dispersed network of settlements based on agriculture, crafts, and trading. Some fourteen hundred village and town sites have been identified so far, with only a few larger cities, notably Harappa with about thirty thousand people and Mohenjo-daro with forty thousand, serving as manufacturing centers and commercial and trading hubs.

Yet there was a higher degree of cultural and technological uniformity throughout this far-flung community than in any other ancient culture. Its standardized system of weights and measures, for instance, was very advanced. As archaeologist Charles Maisels has observed, Harappan/Indus civilization provides a unique example of a self-organized economic system that was not ruled over by a state apparatus, and there is no evidence of centralized planning or administration. The pioneer sociologist Émile Durkheim would have called it a case of "organic solidarity."

Nor is there evidence at any of the population centers of palaces, monuments, treasuries, armories, temples, or even great disparities of wealth. For example, even bronze and copper, among the most valuable resources in early civilizations, were relatively abundant and widely distributed at all levels and in all parts of the commonwealth.

To be sure, some of the pottery, jewelry, and ornaments that have been recovered by archaeologists are of superior quality and were more abundant in some locations, but they seem to be associated with the relative affluence of different trade and crafts guilds and extended family networks rather than with some small, centralized elite. Unlike other ancient civilizations, moreover, the skeletal remains that have been found so far do not reveal sharp differences in nutrition and health within the population.

So why did the Harappan/Indus civilization endure only for about five hundred years (from 2500 to 2000 BC) and then go into a sharp decline? Evidently it was not due to warfare and conquest. The most plausible explanation, which has recently been reinforced with images from the Landsat satellite, is that two of the major tributary rivers, the Sutlej and the Yamuna, where many of the largest settlements were located, radically changed their courses as a result of tectonic movement caused by the upthrusting of the Indian Plate against the Asian Plate. Since the Harappan/Indus people had been able to exploit vast stretches of alluvial floodplains for farming, they had no need for the elaborately engineered canals and irrigation systems found in the Middle East, so they were unable to adapt to the loss of their self-watering soils.

One other likely cause of the decline, proposed by biologist Paul Ewald, was the spread of cholera through the population owing to the practice of piping sewage away from various urban centers and allowing it to contaminate freshwater sources, which other ancient civilizations did not do.[29] In any event, there was a steep population decline. The widely dispersed crafts industries and trading networks shriveled, and the remaining population reverted to living once again in the small, more isolated villages that had characterized the earlier post-Neolithic era.

In the absence of other evidence, Maisels's conclusion seems justified: "I see Indus civilization as an oecumene, or commonwealth, not a state-ordered society. Accordingly, it is the only complex society known to history that truly merits the name of 'civilization' in the proper, non-technical sense. . . . This is the condition of serving the greatest good of the greatest number through advances in knowledge, civility and economic well-being shared by all."

What remains unanswered in the archaeological studies of the Harappan/Indus civilization is how it managed to avoid the violent, hierarchical, and exploitative histories of all the other ancient civilizations, from China to Mesoamerica. I suggest three possible differences that could have made all the difference, to borrow a turn of phrase from anthropologist Gregory Bateson.[30] Harking back to Robert Carneiro's theory about the relation between warfare and environmental circumscription, it is significant that there was a relatively low population density throughout this vast fertile area, along with an abundance of needed resources—from arable land to freshwater, firewood, timber, and metals. Overall, there seems to have been a high level of economic security. It is comparable to the early development of the United States during the nineteenth century, when the government was giving away undeveloped land to the pioneers and when the central government was small and the tax burden was light.

A second major difference in the Harappan/Indus case was that all the other ancient civilizations arose from mergers (mostly by force) among already hierarchical, warring chiefdoms, often in the context of rapidly rising populations with ultimately finite resources (environmental circumscription). Oppressive states were merely elaborations on an already oppressive pattern of political domination. This was not true in the Harappan/Indus case. Finally, there were (apparently) no large, warlike competitor states in the vicinity. External defense against potential enemies was therefore not an imperative that a ruling class might have been tempted to use for its own advantage.

These differences point to some important lessons, I believe. It is paradoxical that the very features that contributed to the emergence of complex societies—agricultural surpluses, larger populations, a division of labor, impersonal economic markets, and the need for a centralized government to defend the state and administer justice—also created conditions that undermined the principles of equality, equity, and reciprocity. Sociologists refer to it as "alienation" and speak of "social distance." In such fractionated and emotionally disconnected social environments, our sense of fairness can be overwhelmed or rationalized away by the compulsions of self-interest and conflicts of interest. And if like-minded people form coalitions, or factions, or parties, or classes, the potential for social conflict is magnified.

In extreme cases, an emotional disengagement sets in and xenophobic (we/they) hostilities can replace any sense of shared purpose and common interests. These extreme cases are all the more likely when there are severe stresses associated with environmental circumscription, or when economic hardship is coupled with extremes of wealth and poverty. The historical record provides endless examples, with ominous implications for our own society. (A landmark study of this dynamic is sociologist Barrington Moore's 1978 book *Injustice: The Social Basis of Obedience and Revolt.*) I will talk further about this social pathology later on.

Finally, the Harappan/Indus civilization also illustrates that there is no inevitable connection between economic and cultural progress and an oppressive, exploitative state apparatus. These are matters of choice. The people responsible for producing the great inequities that were so prevalent in ancient states—and in many modern states as well—could have made more equitable choices. Karl Marx was flatly wrong about any iron laws of history, or the inevitability of exploitation. As James Madison, one of America's Founding Fathers, put it in *The Federalist Papers,* "The latent causes of faction are thus sown in the nature of man."[31] Yes, but those seeds will sprout only when they are planted in fertile soil and well watered.

The story of ancient Athens, the wellspring of Western European civilization, offers a well-documented case study of this historical lesson, as well as providing some insight into the influences that shaped the thinking of one of history's greatest political theorists. For two generations before Plato was born (in about 428 BC), the politics of Athens was dominated by its preeminent leader, Pericles, who orchestrated the prototype for all future golden ages and then presided over its decline. It's a sobering warning for our own time. (See also the comments in the endnote.)[32]

After the Athenian navy under Themistocles had inflicted a crushing defeat on an invading Persian fleet at the battle of Salamis in 480 BC, in one of the greatest of all sea battles (half of the four hundred Persian warships, manned by some forty thousand sailors, were sunk), Athens entered a brief period of military, economic, and political predominance among the city-states of the Greek peninsula. It was hardly a peaceful era, however. The Athenians were confident of their own hegemony, but conflicts with other Greek city-states were endemic.

During these golden years, the Athenian economy thrived. Prosperous farms and small manufacturing sites, many worked by slaves, produced an abundance of material goods. Athens's vital seaport Piraeus—then the foremost in the Mediterranean—swarmed with fishing skiffs and the ubiquitous triremes (galleys with three tiers of oarsmen) that were favored by the Athenian navy. On every tide, a parade of the slower, double-ended merchant sailing vessels fanned out around the Mediterranean carrying exports of such things as olive oil, wine, pottery, honey, and clothing or textiles. In return, Athens imported large quantities of grains, timber, metal ores, papyrus, dried figs, nuts, jewelry, woolen goods—and slaves. A daily gridlock of horses, wagons, and people plied the heavily trafficked, dusty road from Piraeus to the city center and the Agora, or central marketplace. And undergirding all this economic activity was a strong currency system and an elaborate structure of commercial relationships and regulations. As Pericles himself observed: "All the produce of every land comes to Athens."[33]

Athens's material abundance also supported an affluent urban population that probably numbered about one-third of the famous city-state's estimated total of 300,000 inhabitants. Many other wealthy societies, historically, have squandered their surpluses on lavish self-indulgence by the rich and powerful, but Periclean Athens dedicated a share of its material bounty to something more enduring. In effect, Athens used its wealth to produce the great cultural artifacts that we associate with the birthplace of Western civilization. And it was principally Pericles who, with consummate political skills, fostered this cultural flowering. As his biographer, Plutarch, put it: "He did caress the people." Pericles was hardly free from personal faults, but to quote Plutarch again, "He had a noble spirit and a soul that was bent on honour."[34]

Great architectural masterpieces, especially the Parthenon, the Propylaea, and the Temple of Athena Nike, arose on and around the flat-topped hill called the Acropolis (literally "high city"). The sciences flourished, particularly mathematics, physics, biology, and medicine. Writers, poets, painters, and musicians created an abundance of breathtaking works and founded schools dedicated to passing their skills along to others. Live theater, employing the novel idea of using formal plots and characters to convey stories with moral lessons, became a popular pastime. (Some of the plays written by Aeschylus, Sophocles, Euripides, and Aristophanes, among others, still survive.) In those days, too, the wealthy were civic-minded, providing meals for the poor and money for public facilities like baths and hospitals. Wealthy benefactors also subsidized the construction of Athens's crowning architectural achievement, the Acropolis. In other words, Athens enjoyed a relatively harmonious social contract between the rich and the rest of the population.

Equally important, a series of political reforms and innovations over more than a century led to a new model of democratic government, featuring widespread citizen participation. Indeed, the Greeks invented the term *dēmokratia*—rule by the people (*dēmos*). And the centerpiece of Athenian democracy was an open Assembly (*ekklēsia*) that any citizen could attend, although a smaller "Council" of five hundred, chosen each year by lot, was the most influential governing body. Also important was the idea that the law must be sovereign, and that all men must respect the law. Following the reforms engineered by the great lawgiver Solon a century earlier, Athens had evolved a formal system of courts and the novel idea of jury trials, where the court's decisions were rendered by a defendant's fellow citizens.

But perhaps the most distinctive feature of Periclean Athens was its visionary ideal of a Greek (Hellenic) civilization—a commitment to the advancement of knowledge, rational discourse, and ethical conduct in public affairs. This was embodied most especially in Athens's elaborate educational system, which included both general education and a variety of specialized schools. And the capstone of this edifice was the discipline of philosophy: philosophers were among Athens's most respected citizens. Pericles rightly called Athens a "school for all of Greece."[35]

However, this glorious edifice began to crumble in 431 BC when Sparta declared war on Athens (although Athens had provoked it). The disastrous Peloponnesian War was actually not a single war but a series of wars, interrupted by periods of uneasy truce (cold wars). The underlying casus belli were complicated, but Pericles' ambitious moves to expand Athens's external empire (the so-called Delian League) were important factors, along with an overestimate of the ability of its powerful navy. The hostilities between Athens and Sparta continued inconclusively for twenty-seven years.

Early on in the war, Pericles made a fateful decision to withdraw the population into the city proper (fortress Athens) and to abandon the countryside to the superior Peloponnesian alliance—a formidable enemy with a much larger and more disciplined land army. Pericles probably had no realistic choice, but the results were tragic. Athens was largely cut off from its domestic food supply and forced to rely on costly imports by sea. The estates of the wealthy were ravaged, and the angry oligarchs turned in fury against Pericles. Within a year, Pericles was removed from office in a virtual coup and was indicted and fined for embezzlement. However, there was a strong democratic backlash, and he was soon restored to office by a deeply divided electorate.

Then the plague hit Athens. Most likely it had been imported by traders from Carthage. In an overcrowded and stress-filled urban environment like wartime Athens, the results were devastating. The plague started in 430 BC, and within a year perhaps one-third of the entire population had died, including many of Athens's best hoplite soldiers (an elite, body-armored infantry drawn from its best and brightest citizens). Two of Pericles' own sons and a sister died of the plague, and then a grieving Pericles himself succumbed.

No leader of Pericles' stature—much less his military skill in the field—emerged to replace him. After a succession of inconclusive battles and truces, Athens suffered a series of devastating defeats, especially during its misguided invasion of Sicily in 416 BC, where some fifty thousand Athenian soldiers died. The final blow came when the vaunted Athenian fleet was attacked by a brand-new Spartan fleet (built with Persian money) while it was lying at anchor; many of the crews were ashore buying groceries. Only 9 out of 180 Athenian triremes survived. It was perhaps the easiest victory in all of naval warfare until the Japanese air attack on Pearl Harbor in World War II. In April of 404 BC the Spartan admiral Lysander took control of Athens and imposed a reactionary oligarchy along Spartan lines.

This is the Athens that shaped Plato's pessimism. An impoverished, demoralized, angry population became deeply divided politically, and economic tensions ran high. Most of Athens's wealth was now concentrated in the hands of 5 to 10 percent of the population, while 60 to 70 percent lived in more or less severe poverty. Some historians describe the postwar oligarchy (the so-called Thirty Tyrants) as a reign of terror. Eventually the democracy was restored in name, but it really amounted to another radical oligarchy. Political extremists on both sides had become the dominant players in shaping public policy.

It's no wonder Plato chose *not* to use Athens as his model state in his great dialogue *The Republic.* Instead, he went back to the drawing board and tried to design an ideal state that he believed could resolve the fundamental problem of social injustice. Indeed, the little-known and seldom-used subtitle of *The Republic* is *Concerning Justice*. Call it social engineering.[36]

To anticipate my own bottom-line assessment of *The Republic,* Plato had the right diagnosis but the wrong prescription. Yet, to his everlasting credit, he later recognized the flaws in his design and, in his last essay (*The Laws*), outlined what has remained for many subsequent generations—including our own Founding Fathers—a far better solution to the problem of achieving a reasonably just state.[37] This is why it is often said (after the philosopher Alfred North Whitehead) that the history of political theory is merely a footnote to Plato, although it might be more accurate to call it a transgenerational dialogue with Plato. This is also why, in my judgment, *The Republic* and *The Laws*—along with the *Politics* of Aristotle (Plato's most illustrious student and the world's first great polymath)—should be required reading for every modern statesman, although it would do little good in what might be called a Mugabe state, or a Stalinist state, or an ayatollah state for that matter.[38]

At the outset of *The Republic,* Plato raises and rebuts some of the other ideas about politics that were then "in the air" in Athens, many of them attributable to the Sophists. The Sophists were a group of itinerant teachers whose pupils included many of Athens's wealthy aristocrats, who paid generously for being told what they wanted to hear. Among other things, the Sophists taught that all laws are merely social conventions and that each individual has the right to define for himself (or herself) what is right and wrong. For instance, the Sophist Antiphon suggested that some laws may even require us to do what is "unnatural"—helping others. What *is* natural is to pursue your self-interest. Sound familiar?

Later Sophists went even further, arguing that all laws arise from a voluntary compact that can be changed or even subverted if desired. Since inequality is a basic law of nature and we are inherently unequal, justice is whatever the strongest and most powerful can impose on others. Might makes right. Thus the character Thrasymachus in *The Republic* declares that justice is nothing more than "the interest of the stronger." Similarly, the Greek historian Thucydides noted that the democratic Athenians often justified their empire by claiming it was the right of the strong to rule over their weaker inferiors: "The strong do what they can, and the weak suffer what they must. . . . Of the gods we believe by tradition, and of men we know for a fact, that by an irresistible law of Nature, they rule wherever they can."[39] Centuries later, Darwin's political champion—Thomas Henry Huxley (who was referred to as Darwin's bulldog)—characterized it as "tiger-rights."[40] Today we refer to it (jokingly) as the eight-hundred-pound gorilla.

Some of the later Sophists were even more reactionary. They attacked the supposed selfishness of the masses and claimed that democracy allowed the poor to exploit the rich! (Some other later Sophists, who were in the opposite ideological camp, were attacked as political subversives for opposing slavery, claiming it had no objective basis, and especially for their audacity in questioning the subordinate status of women.)

Plato's rebuttal to all these egocentric and individualistic (prelibertarian) arguments proceeded from his core assumptions about the very nature and purpose of human societies. To Plato, justice is not primarily concerned with some higher metaphysics, or a tug-of-war over our rights as individuals. It is concerned with equitable rewards for the proper exercise of our abilities and our calling in a network of interdependent economic relationships. Moreover, and this point is crucial, Plato recognized that fairness also has a floor—a minimum wage, so to speak. Here are Plato's words in *The Republic:*

> If we begin our inquiry by examining the beginning of a city, would that not aid us also in identifying the origins of justice and injustice? . . . A city—or a state—is a response to human needs. No human being is self-sufficient, and all of us have many wants. . . . Since each person has many wants, many partners and purveyors will be required to furnish them. . . . Owing to this interchange of services, a multitude of persons will gather and dwell together in what we have come to call the city or the state. . . . [So] let us construct a city beginning with its origins, keeping in mind that the origin of every real city is human necessity. . . . [However], we are not all

> alike. There is a diversity of talents among men; consequently, one man is best suited to one particular occupation and another to another. . . . We can conclude, then, that production in our city will be more abundant and the products more easily produced and of better quality if each does the work nature [and society] has equipped him to do, at the appropriate time, and is not required to spend time on other occupations. . . . Where, then, do we find justice and injustice? . . . Perhaps they have their origins in the mutual needs of the city's inhabitants.[41]

In other words, Plato correctly identified the underlying purpose of a human society—to provide for the basic survival and reproductive needs of its members—and he fully apprehended one of its fundamental advantages, a division of labor and specialization. (He might also have mentioned the additional benefit of jointly produced collective goods, or public goods.) Plato also specifically characterized a human society as like an organism with many parts (presaging Herbert Spencer's "superorganism" analogy). Most important, Plato grasped the core political challenge—to achieve social justice (fairness) for all the members of the collective survival enterprise. He also defined social justice in a way that has withstood the test of time. In the words of one of Plato's characters, Polymarchus, it involves "giving every man his due."

There have been countless debates through the centuries over what Plato meant by "due." But a commonsense interpretation is that the rewards provided by society should be proportionate to both one's needs and one's contributions, and that the same holds true for crimes and punishments. Plato clearly did not mean equality per se. Rather, he meant an equitable portion in accordance with some criterion of fairness—a fair share. As I noted earlier, Aristotle also used the term proportionate equality. (There is also a voluminous scholarly literature, especially in welfare economics, on Aristotle's related concept of "distributive equity.")

Plato, in his basic recipe for achieving social justice in a complex society, got several ingredients right, I believe. One key ingredient was his insistence that any system of government must seek to be disinterested, or nonpartisan, and should seek to rule in the public interest—in other words, in a manner that is fair to all the classes and interests in a society. A second ingredient was Plato's deep conviction that education is vital to achieving and sustaining social justice. Many of the troubles he saw around him could be traced to unenlightened attitudes and behaviors that were socially destructive. They invited an unbridled selfishness and a wanton disregard for the public interest. "Ignorance is the ruin of states," he believed.

Finally, Plato recognized what many of the Sophists of his day (and many conservative economists and libertarians today) failed to acknowledge. Human societies are not simply a marketplace or an aggregation of isolated individuals seeking to exploit one another or to be free from one another. Humans are fundamentally social animals who are shaped by, and benefit from, participation in the life of the community. So a harmonious society provides important social and psychological benefits as well as economic advantages.

In my opinion, Plato's assumptions represent a sturdy foundation on which to build a political system that embodies social justice. Where Plato took a wrong turn was with his prescription for how to achieve this—his ideal state. What Plato proposed, in a nutshell, is that political power should be entrusted to specially educated and trained "philosopher kings" who would have no personal property, who would live communally as celibates, and who would therefore rule in the public interest. It has often been noted that Plato was not a democrat: he distrusted what has sometimes been referred to as "mobocracy." Thus he preferred an authoritarian model—a benevolent dictatorship. The masses would be ruled over like children by wise, disinterested rulers. Pure reason would prevail over our "animal spirits" and selfish interests.

Some of the problems with this model are obvious. No amount of education can completely erase and rewrite human nature—our evolved "spirits" and "appetites" (to borrow some of Plato's terminology). The landscape of history is littered with political (and religious) gods who, in the end, had feet of clay. Furthermore, some of us are less teachable, more impulsive, and stronger-willed than others. Plato overestimated the power of education to teach self-control. Equally important, a desire for autonomy and personal freedom is also an important part of human nature. Not only may we deeply resent being told precisely how to live our lives, but we may actually know better than any philosopher king (since they could not be omniscient) what we are fitted for and what is best for us—though sometimes our parents, families, friends, and teachers can be helpful. So, at a deep level, Plato's psychology was seriously deficient. (I will have more to say about all of this in the next chapter.)

When Plato set out to write *The Laws,* near the end of his life, his views had drastically changed.[42] He had come around to the view that the next best thing to an all-powerful, all-knowing, and benevolent sovereign was the sovereignty of the law—"a government of laws and not of men," as John Adams, one of our Founding Fathers and our second president, later phrased it.[43] Plato's overall objective remained the same—to secure social justice—but he had come to the view that the law represented a repository of the collective experience and wisdom that had been painstakingly—and sometimes painfully—accumulated by a society over time. At its best, the law represented a body of rules, norms, and practices that were generally accepted as being fair—"reason unaffected by desire," as Aristotle put it in his *Politics.* If everyone was equally subject to the law, and if everyone was treated equitably under the law, this would go a long way toward ensuring a just society. Plato called it "the golden cord of law." In other words, Plato ultimately embraced the great vision of his predecessor, the lawgiver Solon. His thinking had evolved. (See the comparison between Plato and Rand in the endnote.)[44]

However, Plato also recognized that something more is needed to achieve a just society, namely, a set of political institutions that will govern in a way that furthers the underlying purpose of a society and advances social justice. And it is here that Plato made his greatest contribution to political science. Instead of investing power in the hands of philosopher kings, or turning it over to the masses, he proposed that any government is more likely to be just and equitable if it represents an amalgam, or mixture, of the various interests and classes in society. Plato's student Aristotle later elaborated on the idea of a "mixed government" in the *Politics,* suggesting that it should combine democratic (egalitarian) elements, aristocratic elements, and even strong leadership in the form of a "monarchical" element.[45] If all members of a society are fairly represented, this would impel them to work toward compromises that would balance various interests, including the property interests of the rich and the basic needs and wants of the rest of society.

Over the past twenty-five hundred years, evolving human societies have often ignored Plato's (and Aristotle's) seasoned prescriptions, or more often have displayed "ignorance" of them. On the other hand, some societies have been inspired to move toward this model. Western democracies, however imperfect, have been deeply influenced by the teachings of Plato and Aristotle. The mixed forms of government that today we (somewhat imprecisely) call democracies and that employ systems of "checks and balances" resemble in spirit Plato's and Aristotle's vision, despite their flaws. But our progress has been tentative. There have been many setbacks, and there are still many players on the world stage today who don't "get it." As Plato insisted, the key political problem is fairness (equality, equity, and reciprocity). Equitable rewards for merit are justified so long as the basic needs of all the members of society are provided for. That is the very essence of social justice. As Aristotle wisely warned, "Man, when perfected, is the best of all animals but when separated from law and justice is the worst of all."[46]

4 * Fairness and the Science of Human Nature

> Of all the differences between men and the lower animals,
> the moral sense or conscience is by far the most important. . . .
> It is the most noble of all the attributes of man.
>
> CHARLES DARWIN

What Darwin was alluding to here is human nature, the biological substrate that absolutely determines our basic needs and greatly influences our day-to-day wants and preferences—not to mention our moral inclinations. As psychologist David Barash expressed it, our biological natures "whisper" to us in everything we do, and social justice is concerned with how our human natures—lurking just inside our sometimes elaborate cultural outer garments—interact with the other human natures we encounter.

We may not be accustomed to thinking about ourselves—or one another—in such a biological way. Despite a sea change in attitudes in recent years, there are still many social theorists who remain hostile to the very idea that our minds are not free of our bodily urges and cannot be endlessly manipulated by cultural influences. They still cling to the "blank slate" (tabula rasa) model and to Émile Durkheim's famous dictum, "Every time a social phenomenon is directly explained in terms of a psychological phenomenon, we may be sure that the explanation is false. . . . The determining cause of a social fact should be sought among the social facts preceding it and not among the states of individual consciousness."[1]

Durkheim's theoretical stance was flatly wrong. Human nature is one of the costars in the human drama. This is becoming ever more apparent as the burgeoning science of human nature acquires a deeper understanding of what goes on inside the "black box"—the enormously complex system we call the human mind. Moreover, the assumptions we make about human nature matter a great deal. As Steven Pinker points out in his best-selling tour de force on our modern-day misconceptions (and biases) about human nature, appropriately titled *The Blank Slate,*

> Everyone has a theory of human nature. . . . Our theory of human nature is the wellspring of much of our lives. We consult it when we want to persuade or threaten, inform or deceive. It advises us on how to nurture our marriages, bring up our children, and control our own behavior. Its assumptions about learning drive our educational policy; its assumptions about motivation drive our policies on economics, law, and crime. And because it delineates what people can achieve easily, what they can achieve only with sacrifice or pain, and what they cannot achieve at all, it affects our values. . . . Rival theories of human nature are entwined in different ways of life and different political systems, and have been a source of much conflict over the course of history.[2]

Indeed, our theories of human nature have been co-conspirators in some of history's most horrendous human tragedies. The great twentieth-century economist John Maynard Keynes may have had Adolf Hitler in mind when he commented that "mad men in authority, hearing voices in the air, are distilling their frenzy from some academic scribbler of a few years back."[3] Accordingly, a scientific approach to human nature is where true wisdom can be found.

Let's start with some perspective on how human nature has been characterized in the philosophical literature over the centuries. The problem is that our progress in trying to understand the often subtle influence our biological nature exerts over us has been painfully slow. For thousands of years, our philosophers and savants have had to rely on various combinations of folk wisdom, the promptings of their immediate cultural and political environments, and their personal predilections and experiences. Call it hearsay—a muddle of often insightful but also sharply conflicting opinions. Here is just a sampler.

Plato, in *The Republic,* divided what he called the "soul" into three elements: "appetitive" (nutrition, sex, etc.), "spirited" (emotions, ambition, competitive urges, etc.), and a rational, reasoning element that he viewed as the primary function of the brain. These three elements can be either in harmony or at war with one another, Plato argued, so it is imperative for the rational element in each of us to exercise control over the urges and impulses that arise from below.

Human societies confront a similar challenge, Plato believed, and in *The Republic* he envisioned that his ideal state would comprise three social classes that roughly correspond to the three elements of our individual souls. If humans with varying talents and temperaments are properly educated and trained—from manual laborers to soldiers to philosopher kings—and are then given the work they are suited for, the result would be a harmonious and just society, which Plato characterized as a sort of moral organism. (Again, Plato was perhaps the first great philosopher to view society as an organism.) Justice, therefore, accords with our character as social animals; as Plato expressed it, we are "just by nature." And the key to achieving justice in any society is to provide for rule by disinterested wise men.

To his credit, Plato (and Aristotle) also grasped the fundamental duality of human nature and the tensions that exist between our egoistic and our social tendencies. We are, in fact, a tangle of conflicting impulses. We are at times fierce competitors and at other times enthusiastic cooperators; we are both eager for social approval and deeply offended by personal slights; our emotions range from genuine sympathy to righteous anger and from profound love to raging hate. Plato and Aristotle well understood our contradictory natures and appreciated the necessity for the controls (and self-controls) that education and a system of law and order can provide. Many generations of their successors in the tradition of discourse have echoed this dualistic view, from John of Salisbury to Marsilio of Padua, Saint Thomas Aquinas, Jean Bodin, Edmund Burke, and many modern-day liberal theorists.

The radically individualistic, self-centered model of human nature that was promoted by the Sophists has also had a continuing voice in the transgenerational dialogue. For instance, the fourth-century BC Greek philosopher Epicurus, writing at a time when a dispirited Greece had become a vassal of the Roman Empire, adopted the Sophist argument that human nature is governed by self-interest and that the self desires above all to obtain happiness and avoid pain and stress. There are therefore no intrinsic moral rules, only variable customs. And any social compact exists only as a means for satisfying our private desires. Epicurus tells us, "There never was an absolute justice but only a convention made in mutual intercourse, in whatever region, from time to time, providing against the infliction or suffering of harm."[4]

The Sophist and Epicurean models of human nature reappeared again in the late eighteenth century and the nineteenth century in the writings of the so-called utilitarian philosophers and early classical economists—most notably in the work of Jeremy Bentham, John Stuart Mill, and David Ricardo. As Bentham expressed it, "Nature has placed mankind under the governance of two sovereign masters, *pain* and *pleasure*. It is for them alone to point out what we ought to do, as well as to determine what we shall do." Therefore it is "the greatest happiness of the greatest number that is the measure of right and wrong."[5] As various critics have pointed out, Bentham's principle is blind to the concepts of fairness and ethics. Thus, if torturing a few will benefit many others, so be it.

Then there was that notorious Machiavellian, Niccolò di Bernardo dei

Machiavelli, who was shockingly cynical about politics and denigrated any moral aspirations that did not take into account the brute facts of human nature as he saw them. "One can say this in general of men: they are ungrateful, disloyal, insincere, deceitful, timid of danger and avid of profit. . . . Love is a bond of obligation which these miserable creatures break whenever it suits them to do so; but fear holds them fast by a dread of punishment that never passes."[6] In short, Machiavelli was a disillusioned moralist with a political agenda.

Machiavelli is commonly accused of advocating the use of ruthless, amoral, and deceptive behavior in pursuit of personal self-interest, but this is flagrantly untrue. He was, in fact, a fervent patriot who dreamed of inspiring a prince, such as the brilliant soldier-churchman Cesare Borgia, to take up the cause of consolidating Italy's deeply divided, corrupt, and warring states into a unified nation. He hoped to find a "Machiavellian" philosopher king who could bring about law and order and achieve political stability in the anarchic political environment of the Italian peninsula in his day.

Accordingly, Machiavelli's most famous tract, *The Prince,* was really a handbook for how a nation builder in sixteenth century Italy would have to proceed in the political world of his time and place. In effect, Machiavelli argued that the end justifies the means, and that despicable means were indispensable. As he bluntly advised, "It is well that, when the act accuses him, the result should excuse him; and when the result is good [serving beneficent purposes] . . . it will always absolve him from blame. . . . For, all things considered, it will be found that some things that seem like virtue will lead you to ruin if you follow them; whilst others, that apparently are vices, will, if followed, result in your safety and well-being."[7]

Thomas Hobbes, the controversial English philosopher (and tutor to the very wealthy), lived during a similarly turbulent period around the time of the English civil wars in the mid-seventeenth century. Although he aspired to a scientific approach to politics, he produced instead an extension of Machiavelli's dark view of human nature. In his great work *Leviathan* (the Hebrew word for sea monster) in 1651, he wrote: "I put for a general inclination of all mankind, a perpetual and restless desire of power after power that ceaseth only in death. And the cause of this is not always that a man hopes for a more intensive delight than he has already attained to . . . but because he cannot assure the power and means to live well . . . without the acquisition of more."[8] (Actually, Hobbes's reasoning was a bit more complicated than this. He gave three reasons why humans fight with one another: competition over material goods, general distrust, and power seeking.)

To Hobbes, therefore, there is no such thing as justice. The state of nature is "a war of every man against every man." And the basic law of nature is "only that to be every man's what he can get; and for so long as he can keep it." For the great majority of humankind, life is "solitary, poor, nasty, brutish, and short."[9] On the other hand, our desire for power is equaled by our fear of death. Therefore governments are created to provide for our self-preservation—that is, security against our neighbors, who are roughly of equal strength. Because we are always in mortal danger, Hobbes asserted, only a monarch with unrestricted power can guarantee peace and civil order. As Hobbes explained it, "Such gentler virtues as justice, equity, mercy, and in sum, *doing unto others as we would be done to,* without terror to cause them to be observed, are contrary to our natural passions. . . . Covenants without the sword are but words, and of no strength to secure a man at all. . . . The bonds of words are too weak to bridle men's ambition, avarice, anger, and other passions, without the fear of some coercive power."[10]

Hobbes's sovereign is thus a kind of flipside to Plato's philosopher king. Instead of an all-powerful but benevolent dispenser of social justice, Hobbes proposed an all-powerful "policeman" who would be able to keep the peace.

Hobbes's conclusions have at least two grains of truth. One is that there may indeed be the need for a strong peacekeeper and a unifying force when a society has deep political divisions and social turmoil such as Hobbes witnessed. The other is that, even in more peaceful times, policing is required to deter and punish the unavoidable cheaters and free riders (see below). Yet Hobbes's underlying assumptions about human nature were a bit too jaundiced. To be sure, "self-preservation" is a fundamental human concern and could even be viewed as roughly synonymous with Darwin's "struggle for existence." But Hobbes's psychology was one-sided. And so was his hypothesis about the state of nature—as we now know.

More important, Hobbes seemed oblivious to the Roman poet Juvenal's famous question, attributed to Socrates in *The Republic*: "Who will guard the guardians?" (*Quis custodiet ipsos custodes,* as it is rendered in Latin). As the nineteenth-century English historian Lord Acton famously put it, "Power tends to corrupt, and absolute power corrupts absolutely."[11] Or in the words of the former United States secretary of state Henry Kissinger, power is "the greatest aphrodisiac in the world."[12] Sadly enough, a great many societies have had to learn this lesson the hard way.

John Locke, by contrast, was a person of a very different temperament who enjoyed a far more benign and stable political environment in England toward the end of seventeenth century. A moderate, ethically minded Christian who was also a practicing physician, Locke took up the cause of the rising commercial classes. He was especially concerned about how to constrain and limit the power of the monarchy and the church, and in his *Two Treatises of Government* he specifically refuted Hobbes's absolutism.[13]

Locke's seminal work was based on a very different set of assumptions about human nature—that all men started out free and equal in "the state of nature" and willingly cooperated for their mutual advantage. So when they came together voluntarily to form a more advantageous social contract, they retained their fundamental rights, especially their property rights. Among other things, Locke argued that societies do not create property. Each of us does so when we "mix" our labor with an object or a piece of land.

Locke was also the first "modern" theorist to articulate the concept of self-evident natural rights, including "life, liberty, and estate." (The American Founding Fathers, concerned that the right to an estate, or property, might not be so self-evident to many people, substituted "the pursuit of happiness" in their rendering of Locke's idea.) Furthermore, Locke asserted, the social contract implies that governments exist to protect our inherent rights. So we can see in Locke's theory of human nature the inspiration for the American Declaration of Independence and Bill of Rights, the French Declaration of the Rights of Man, the United Nations' Universal Declaration of Human Rights, and modern libertarianism, among other things. (On libertarianism, see the endnote.)[14]

However, our rights are not absolute, according to Locke. They are limited by the caveat that they must not interfere with the rights of others. In the same vein, Adam Smith, in his less well-known early work *The Theory of Moral Sentiments,* stressed the importance of doing justice, which he defined as not doing "injury" to others. "There can be no proper motive for hurting our neighbor."[15] Jeremy Bentham also qualified his pain-pleasure principle by conceding that our freedom must be constrained by the rule that it "affects the interests of no other persons" besides the actor.[16] Libertarians, likewise, generally aver that the exercise of our rights must not cause "harm" to anyone else. As we shall see, these qualifiers represent a huge loophole that, properly interpreted, can be used to turn Locke, Bentham, Nozick, and Rand (among others) on their heads. Or, better said, it can be used to turn them right side up, as I will seek to do in the chapters ahead.

Before we move on to the science of human nature and the biological roots of fairness, there is one other thread in political theory that I should mention briefly. Egalitarian socialism, or the idea of individual equality in a community of equals, can be traced back to some of the radical Sophists and early Christians, but the modern-day fountainhead for this theme (to borrow a term from Ayn Rand) is Jean-Jacques Rousseau. The son of a watchmaker in Geneva, Switzerland, Rousseau had a troubled and emotionally unstable life and stormy relationships with some of his peers. Yet he was a prolific novelist, essayist, and musician who also enjoyed the patronage of wealthy benefactors at various times—which is curious in view of his outrage over the extremes of wealth and mass poverty that he saw all around him. "Man is born free, but everywhere he is in chains," Rousseau famously declaimed.[17]

Rousseau believed passionately that humans are naturally good and cooperative and that they are inherently social beings who are suited by nature for living in communal (and moral) societies. However, Rousseau charged that existing societies had become corrupted by the unbridled pursuit of self-interest, and he specifically rejected the idea that any summation of private interests could be equated with the public good. To Rousseau, the community as a whole—as a moral being—has a higher claim than our individual rights. "The right which each individual has to his own estate is always subordinate to the right which the community has over all." Indeed, "The social order is a sacred right which is the basis of all other rights."[18]

Rousseau concluded that existing societies, which had become the captives of a decadent nobility, must be reconstituted as communities of equals who are subject to the "general will"—a vague and much-debated concept that some theorists have interpreted more benignly as being equivalent to the "public interest" or the "general welfare" and others have viewed as a license for tyranny. Unfortunately, Rousseau encouraged the latter interpretation with his assertions that "the social contract gives the body politic absolute power over its members," and that if anyone resists the general will, "he will be forced to be free."[19] During the French Revolution Robespierre did just that. He justified the actions of his Jacobins with the assertion that "our will is the general will. . . . They say that terrorism is the resort of despotic government. . . . The government of the Revolution is the despotism of liberty against tyranny."[20]

Rousseau certainly didn't intend to encourage an authoritarian interpretation of his work. His vision of the social contract as articulated in his most famous essay, *Du contrat social* in 1762, referred to the sharing of our lives in a community with common interests, a common language and culture, and common property. He saw our social needs as being just as natural as our individual self-interest. And while he was not opposed to private property, he rejected the idea that it constitutes an inherent "natural right." To Rousseau all rights, including property, exist within the community, which has a higher moral claim. He conceived of the general will as a superior claim on behalf of the community as a whole over any private rights.

Needless to say, Rousseau's conclusions had revolutionary implications. It's not surprising that his books have sometimes been banned—or burned. He also provided the conceptual and moral foundation for Karl Marx's vision of a communist society and for the brutally effective rationalizations used by Lenin, Stalin, and Mao during the two great political revolutions of the twentieth century. On the other hand, Rousseau also presaged an assortment of more temperate, more moderate social reformers, from T. H. Green and the Oxford Idealists to the Fabian socialists, socialist political parties in various countries, and more recently, the self-defined Communitarians and their sympathizers, like Amitai Etzioni, Michael Walzer, Charles Taylor, Alasdair MacIntyre, Robert Bellah, Jean Bethke Elshtain, and others.[21]

How, then, do we begin to reconcile these three very different views of human nature and society (organismic, individualistic, and socialistic)? The answer, in short, is that each of them provides some partial insights; each has a different perspective on the same human nature as it has expressed itself in different historical circumstances and cultural arenas. Some theorists have focused on the way we are embedded in and shaped by the various superorganisms that we depend upon ultimately for the satisfaction of our basic needs and wants, and that also influence our social behavior. Other theorists have focused on our individual personalities and our personal needs, desires, and agendas. Still others have focused on the character of our relationships with each other and with the state. To use some social science jargon, they represent both "macro" and "micro" views of humankind.[22]

Though there are some obvious disagreements among them, these three broad perspectives are not as contradictory as they may seem. Each one stresses a different side of an enormously complicated animal in its many different social and political habitats. An appropriate metaphor might be the much-used (apocryphal) story about the blind men and the elephant. Five blind men who have had no previous experience with an elephant are allowed to touch one in a zoo, after which they variously describe it as being like a wall, a tree, a rope, a snake, and a fan. Thanks to the emerging science of human nature, the blinders are beginning to come off, and we can now see the entire elephant more clearly. It appears the elephant is more complex, many-faceted, and paradoxical than many of our philosophers have imagined.

Human nature is rooted in our prehistory as a species: we were shaped in the pressure cooker of human evolution. So let's briefly recall the plausibility scenario discussed in the last chapter. The origins of human societies trace back to a pattern of joint ventures—limited cooperative efforts—

among small groups of close kin and (most likely) some non-kin as well to cope with the rigors of survival and reproduction in a resource-rich but challenging environment. Survival and reproduction were the common objectives, and our ancestral social contract involved a collective survival enterprise. Our ancestors, like ourselves, benefited from close cooperation with others in providing for their basic survival needs.

Plato and Aristotle certainly got this part of the story right, and they also understood the synergies that derived from our progressive specialization of roles and the division of labor. In the terminology of economics, there were both "public goods" that benefited all the participants (from group defense to collective foraging, joint decision making, etc.) and what I refer to as "corporate goods"—cooperatively produced or procured goods that can be divided up and distributed in more (or less) equitable ways. Charles Darwin, in his treatise *The Descent of Man,* provided us with a vivid account of the cooperative aspect of the natural world:

> Animals of many kinds are social; we find even distinct species living together; for example, some American monkeys; and united flocks of rooks, jackdaws and starlings. . . . The most common mutual service in the higher animals is to warn one another of danger by the united senses of all. . . . Social animals perform many little services for each other; horses nibble and cows lick each other for external parasites. . . . Animals also render more important services to one another; thus wolves and some other beasts of prey hunt in packs, and aid one another in attacking their victims. Pelicans fish in concert. The Hamadryas baboons turn over stones to find insects, etc.; and when they come to a large one, as many as can stand around, turn it over together and share the booty. Social animals mutually defend each other. Bull bisons in North America, when there is danger, drive the cows and calves into the middle of the herd, while they defend the outside.[23]

Accordingly, there are a number of important features that we can reasonably infer about our evolving ancestors. The groups that were most successful over time developed a variety of social facilitators and social controls that mitigated and contained, but certainly didn't eliminate, the typical (mostly male) primate tendencies toward individual dominance competition and interpersonal rivalry, as I noted earlier. All members had a strong interest in the preservation and well-being of the group and used various means to maintain harmony and good relationships—just as our close cousins among the chimpanzees and bonobos (the so-called pygmy chimpanzees) are able to do quite skillfully to this day.[24] Yet, at the same time, our ancestors had a strong antipathy toward competing groups and competing species. Group defense (and offense) was of primary importance.

Another important feature of these groups was egalitarian food sharing, a behavior pattern that has been observed in many other socially organized species as well. At what point these groups began to exhibit a broader sense of fairness and equity—a more expansive predisposition to assist others and to accommodate to the needs of the group as a whole—is, of course, impossible to say. But, as we shall see, there is now much evidence in modern-day humankind that this did in fact occur and that it is one of the hallmarks of human nature.

For the Darwinian evolutionary biologists of the past half-century, this seemingly altruistic facet of human nature presented an insurmountable theoretical problem. In fact, some of the leading evolutionary theorists, like William Hamilton, George C. Williams, and Richard Dawkins, have considered human morality to be an inexplicable anomaly—something that developed in spite of our innate selfishness and egoism. As Dawkins put it in *The Selfish Gene,* "A predominant quality to be expected in a successful gene is ruthless selfishness. . . . I think 'nature red in tooth and claw' sums up our modern understanding of natural selection admirably."[25]

Dawkins was seconded by Williams: "I account for morality as an accidental capability produced, in its boundless stupidity, by a biological process [human evolution, presumably] that is normally opposed to the expression of such a capability."[26] On the other hand, biologist Richard Alexander, in his pathbreaking 1987 volume *The Biology of Moral Systems,* reasoned that our moral behaviors are an evolved outgrowth, or spillover, from altruistic behaviors toward close family members—"kin selection," in evolutionary terminology.[27] In other words, they were not really altruistic behaviors because they ultimately served the genes of close relatives.

However, it was Darwin himself who may have provided the definitive resolution to this evolutionary conundrum. What he proposed in *The Descent of Man* was a tripartite selection process in human evolution that today goes under the heading of multilevel selection theory. As Darwin appreciated, natural selection can occur within nested evolutionary units, from individuals to families, organized groups, larger populations, and even symbiotic partnerships between different species. Furthermore, all these selection processes can be going on simultaneously, so that the outcomes are a kind of vector sum of various selective biases. In the evolution of humankind, in particular, Darwin thought that selection between hominid groups was of great importance. He saw no contradiction with subordinate processes of competition and selection among individuals and families. One can quibble with some of his details, but it now seems evident that Darwin was correct about the essential dynamic:

> In the first place, as the reasoning powers and foresight of the members became improved, each man would soon learn that if he aided his fellow-men, he would commonly receive aid in return. From this low motive he might acquire the habit of aiding his fellows. And the habit of performing benevolent actions certainly strengthens the feelings of sympathy which gives first impulse to benevolent actions. . . . But another and much more powerful stimulus to the development of the social virtues is afforded by the praise and blame of our fellow-men . . . and this instinct no doubt was originally acquired, like all other social instincts, through natural selection.[28]

Neo-Darwinian evolutionary biologists in recent decades have been so implacably hostile to the idea of group selection that it was politically incorrect among evolutionists even to discuss such a heretical idea. One theorist went so far as to claim that in this respect Darwin was misguided: he "let us down." The verdict pronounced by George Williams in his influential 1966 critique of evolutionary theory, *Adaptation and Natural Selection,* was viewed by many as an evolutionary New Testament. Natural selection between groups as a source of change, he declared, was "impotent" and "not an appreciable factor in evolution."[29]

But times have changed. Thanks to a rethinking of this issue and the dogged efforts of a new generation of evolutionary theorists, including David Sloan Wilson, Edward O. Wilson, and various colleagues, group selection theory has recently regained respectability, if not universal acceptance. Although it remains controversial, it is now generally conceded that human evolution is perhaps a likely example of group selection. So Darwin may have been right about human evolution after all. As he observed,

> All that we know about savages, or may infer from their traditions and old monuments, the history of which is quite forgotten by the present inhabitants, shows that from the remotest times successful tribes have supplanted other tribes. . . . [Moreover,] natural selection, arising from the competition of tribe with tribe . . . would, under favorable conditions, have suffered to raise man to his high position.[30]

Accordingly, Darwin reasoned that our social and moral attributes were of the utmost importance to our evolutionary success. As he put it, the tribes that were the most highly endowed with "patriotism, fidelity, obedience, courage, and sympathy, were always ready to aid one another, and to sacrifice themselves for the common good" would have had a decisive competitive advantage:

> Selfish and contentious people will not cohere, and without coherence nothing can be effected. A tribe rich in the above qualities would spread and be victorious over other tribes; but in the course of time it would, judging from all past history, be in its turn overcome by some other tribe still more highly endowed. Thus the social and moral qualities would slowly tend to advance and be diffused throughout the world.[31]

Again, one can find fault with some of the details, and certainly with some of Darwin's terminology. For instance, the Victorian conceit that simple hunter-gather societies were "savages" is no longer countenanced. Alas, savagery is possible even in "civilized" societies. But, more to the point, there is reason to qualify and modify Darwin's scenario. The relative importance of differential survival among ancestral hominids as a result of direct violent conflict is very uncertain. It's more likely that primitive warfare was episodic. As I suggested earlier, differential survival is also likely to have been influenced by varying degrees of success in coping with other predator species, as well as in dealing with ecological challenges of various kinds, and maybe even differences in group knowledge and decision-making skills (such as not being in the wrong place when there is a flash flood, or a landslide, or a brush fire). Indeed, differential access to needed resources and territories was also a very important factor. No doubt the dynamics of human evolution were much more complex than we can imagine.

Nevertheless, group selection between socially organized and closely cooperating groups seems likely to have provided a favorable selection regime for the emergence of our unique wardrobe of social and moral traits. A short listing of these would include our close emotional identification with and loyalty to our own "tribe," our eagerness to fit in and accommodate to social norms, our readiness to display public symbols of group membership and patriotism, our concern for the well-being of the group as a whole (the "public interest," the "common good," and, yes, the "general will"), our keen interest in the lives of other people, our readiness to form social attachments and to engage in cooperative activities, our ability to channel and contain disruptive displays of antisocial behavior, our sensitivity to the "praise and blame" of those around us, especially those we care most about, and not least, our altruism, our moral impulses, and our sense of justice and fairness—just as Darwin proposed. Needless to say, these (and other) social traits have not replaced our innate egoism or our competitive impulses, or (for many of us) our striving for power and dominance. Machiavelli and Hobbes certainly got that part right, though they saw it at its anarchic worst.

Beyond this plausibility argument about events and processes that occurred long ago in the deep mists of our evolutionary prehistory, we can test this scenario against what the burgeoning science of human nature has been learning about our species. In fact, there are perhaps twenty distinct research domains (depending on how you count), spanning at least ten academic disciplines, all of which have something to say about this issue. These research domains include animal behavior studies (especially with primates), archaeology, anthropology, behavior genetics, the brain sciences, the cognitive sciences, child development, evolutionary psychology, sociology, political science, and behavioral and experimental economics. Here (briefly) are some highlights from six of these domains:

Animal Behavior

The extensive research in animal behavior, including especially studies of nonhuman primates both in the field and in various laboratory settings, provides some of the most compelling evidence of all. Our (relatively) small-brained cousin species display many of the same traits, at least in rudimentary form, that are widely observed in humans. As primatologist Frans de Waal has documented in several important books—with suggestive titles like *Chimpanzee Politics, Good Natured, Primates and Philosophers,* and *Our Inner Ape*—chimpanzees and bonobos in particular have very complex, though differing, social and political lives.[32]

De Waal has been criticized in some quarters for "anthropomorphism"—attributing humanlike motivations, emotions, and cognitive abilities to our close relatives, as though very different psychological processes must be responsible for producing the similar results. De Waal counters this objection with the argument that the principle of "evolutionary parsimony" (the simplest evolutionary explanation) dictates otherwise. It makes no sense to think that similar traits in closely related species could be based on fundamentally different biological mechanisms—unless, of course, you want to introduce the alternative of divine intervention and human exceptionalism. It is far more likely that these traits are homologous.

In any case, nonhuman primates display many traits that are also commonplace among humans: reciprocity (especially food sharing), a capacity for empathy and consolation, conflict resolution techniques, community concern, deterrents and punishments for cheating, leadership behavior, and what animal researchers refer to as a "control role," as well as coalition building and dominance competition (especially among the males), and a rudimentary form of "justice" in the form of a tit-for-tat for favors and transgressions. Food sharing is especially common in the primates and has been observed even among unrelated animals (beyond a nepotistic "kin selection" motive). It is found in chimpanzees, bonobos, baboons, siamangs, orangutans, and capuchin monkeys, among others. (As an aside, food sharing may also be used in primate societies as a "bribe"—for instance, when males are building "political" coalitions.)[33]

To test for reciprocity behaviors in capuchin monkeys, de Waal some years ago conducted a decisive laboratory experiment. Here is his description of it:

> Adult capuchins were separated into pairs and placed into a test chamber divided into two sections by a [wide] mesh partition. One capuchin was allowed continuous access to a bucket of attractive food. The individual with access to the food was free to monopolize all of it or could move close to the mesh and share actively or passively by allowing his counterpart access to pieces. . . . The situation was then reversed so that the second individual had access to the attractive food. . . . The rate of transfer between pairs of adult female capuchins was found to be reciprocal, while the rate of transfer between the adult males was not. Males, however, were less discriminating than females in terms of with whom they shared, and more generous in the amount of food they shared.[34]

In short, capuchin monkeys are inveterate reciprocators (especially the females). As de Waal points out, these results were not anomalous. They have also been frequently observed at research stations like the Yerkes Regional Primate Center in Georgia (where de Waal is located), as well as in the wild. A sense of fairness may be a deeply rooted primate trait.

There is also a phenomenon in nature, as in human societies, that biologist Robert Trivers dubbed "reciprocal altruism."[35] One of the most dramatic (and well-documented) examples involves vampire bats, which subsist entirely on animal blood (though not human blood, as a rule). In a classic set of field studies over a ten-year period, biologist Gerald Wilkinson found that, when vampire bats return to their communal nests from a successful night's foraging, they frequently regurgitate blood and share it with nest-mates, including even nonrelatives.[36]

The reason, it turns out, is that blood sharing greatly improves each bat's chances of survival. A bat that fails to feed for two nights in a row will die. (No wonder they're so bloodthirsty.) Wilkinson showed that the blood donors are typically sharing their surpluses and, in so doing, are saving unsuccessful foragers that are close to starvation. So the costs are relatively low and the benefits are relatively high. Since no bat can be certain of success on any given night, it is likely that the donor will itself eventually need help from some nest-mate. In effect, the vampire bats have created a kind of mutual insurance system.

Anthropology

The Human Relations Area Files (HRAF) at Yale University contain a priceless trove of data from literally thousands of field studies by anthropologists (and now archaeologists as well) in every corner and crevice of the world. Over the sixty years of its existence, the HRAF (which is actually consortium of more than three hundred academic institutions worldwide) has assembled a database that encompasses about four hundred cultures, both past and present, with information indexed down to the paragraph level. What this rich research repository overwhelmingly confirms is that there is indeed a universal human nature—just what one would expect if we all shared common ancestors despite our obvious though superficial cultural differences.

Earlier I mentioned Christopher Boehm's study with a screened sample of forty-eight hunter-gatherer societies, where he was able to document a pattern of egalitarian sharing and where dominance behaviors were actively resisted by coalitions of other group members. He referred to it as a "reverse dominance hierarchy." Some years ago another anthropologist, Donald Brown, conducted a much more extensive study in which he assembled a comprehensive list of human universals (the title of his landmark 1991 book).[37] To almost everyone's surprise at the time, because anthropologists had been so fixated on the differences between cultures, Brown found literally hundreds of cultural universals. The commonalities went so far beyond coincidence, or being explainable as independent cultural inventions, that the revelation has transformed our view of human nature. (In a similar vein, the human ethologist Irenäus Eibl-Eibesfeldt has painstakingly documented the universality of many human facial expressions and gestures, confirming the pioneering work of Charles Darwin himself in his less well-known study *The Expression of the Emotions in Man and Animals.*)[38]

Many of the common traits that Brown documented had traditionally been treated as idiosyncratic—body adornment, color classifications, play, music, dance, hair styling, kinship classification systems, mourning rituals, decorative art, and others. But many other traits that Brown examined go much deeper into our emotional repertoire—romantic love, jealousy, shame, empathy, retaliation and revenge, male aggressiveness, and social conflict, among others. For our purpose, however, the most important commonalities Brown identified were those that affect our social, economic, and political relationships—widespread cooperation, collective decision making, leadership, political coalitions, inequalities in prestige and influence, resistance to power and dominance (just as Boehm had found), formal rights and obligations (rules and laws), governance processes, social sanctions, and, yes, reciprocity and a sense of fairness, or equity. (Among the noteworthy sex differences—let's be honest—males are much more prone to cheating and disruptive behaviors and are far more likely to engage in coalitional violence.)

The latter finding is especially significant in light of anthropologist Carol Ember's studies of collective violence between human societies.[39] She documented that some 90 percent of all the hunter-gatherer societies represented in the HRAF database had engaged in warfare (violent conflicts) with other groups, 64 percent of them within the previous two years. In a similar study with a smaller sample of ninety-nine groups of hunter-gatherers, anthropologist William Divale found that sixty-eight were reported to be currently involved in violent conflicts and twenty more had been within the past twenty-five years at most.[40] So what the HRAF database reveals about our heritage is a strong pattern of internal order, stability, and equity coupled with xenophobia and rampant hostility toward other groups. This is precisely what we would expect from our understanding of human evolution (as described in chapter 3).

Equally important, however, is a phenomenon that has been obscured in the search for broad cultural commonalities: the many ways a given culture can also engage in strictly harmful and even self-destructive cultural practices. These were described by anthropologist Robert Edgerton in a global survey that he titled *Sick Societies: Challenging the Myth of Primitive Harmony*.[41] As Edgerton put it (paraphrasing a line from George Orwell's famous novel *Animal Farm*), "All societies are sick, but some are sicker than others." Even though a society as a whole may be reasonably healthy and well adapted, there are always likely to be some practices that are harmful to some or all of the population, ranging from wife beating to witchcraft, human sacrifice, lethal competition for women, rape, feuding and revenge, suicide, slavery, drug addiction, smoking, extreme food practices, and many more. (I'll come back to this point later on.)

Another problem highlighted by Edgerton is what happens to a human society that is under severe stress, such as Athens after the Peloponnesian War, or England during the civil war period in the seventeenth century, or Machiavelli's Italy. The various ways societies respond to severe stresses of various kinds were vividly illustrated in two in-depth field studies by anthropologist Colin Turnbull that became best-selling books.

In his first book, *The Forest People* (in 1961), Turnbull detailed the lifestyle of the Ba Mbuti pygmies in what was then the Belgian Congo (now Zaire). The Ba Mbuti had responded to the encroachments of civilization by returning to live in the Ituri Forest, where Turnbull found a highly cooperative and generally happy society consisting of a number of interrelated families that relied on joint hunting and gathering, broad sharing, and a cooperative work pattern.[42] The Ba Mbuti also maintained a flexible political hierarchy, engaged in active trading with other groups, had an almost religious reverence for the forest ("when the forest dies, we die," they were fond of saying) as well as for their "sacred fire," which they carried with them in the form of embers wrapped in fire-resistant leaves when they migrated to new campsites. Group singing was also an important part of their culture.

Turnbull then spent several years studying the Ik in northern Uganda, and he described what he found in his 1972 book *The Mountain People*.[43] In contrast to the more or less successful lifestyle changes among hunter-gathers like the Ba Mbuti, the Ik had been forced by the Ugandan government into a settled farming way of life in a mountainous area with poor soil fertility and frequent droughts. Also, they were prohibited from pursuing their traditional pattern of hunting and gathering in what had been declared a national park.

Under these severe political and ecological pressures, Ik society unraveled and became a libertarian nightmare—much closer to Hobbes than to Locke, and a universe away from Rousseau. Every man for himself meant just that—no help even for close neighbors or relatives, incessant conflict, wife beating, stealing of food, hostility toward those in need, rejection of the old or infirm, and even the abandonment of children. Turnbull concluded, "It is certainly difficult . . . to establish any rules of conduct that could be called social, the prime maxim of all Ik being that each man should do what he wants to do, [and] that he should do anything else only if he is forced to." (Later on I'll return to how societies react to severe stress.)

Behavioral Genetics

This important interdiscipline, which melds psychology and various biological sciences, has had a checkered history.[44] In its early days, in the late nineteenth and early twentieth centuries, its progress as a scientific discipline was hampered by the so-called eugenics movement—public campaigns in various countries that were based on a misreading of Darwin's theory and with no concrete evidence to speak of—to sterilize those deemed "unfit," mainly among the poor and mentally ill. The most flagrant example was the wholesale extermination program of the Nazis during World War II. Eugenics rightfully became a pariah after the Holocaust. Likewise, in the 1970s and 1980s, there was virulent controversy in our country over racial differences in intelligence that were purported to be genetically determined. More recently, a similar debate erupted at Harvard University over supposed sex differences in mental abilities.

Nevertheless, during the past century the science of behavioral genetics has made impressive though mostly quiet progress in illuminating the role of the genes in shaping our behavioral propensities—human nature—just as our genes shape the rest of our physiology.[45] Behavioral geneticists have shown beyond any reasonable—or even unreasonable—doubt that our genes play a powerful role not only in the overwhelming majority of our behavioral commonalities but also in our many subtle but important behavioral differences. However, anyone who would like to use these findings as a political bludgeon should be aware that (1) genetic similarities vastly outweigh any differences, (2) environmental and cultural influences are equally powerful and are much more easily modified in socially desired ways, (3) all our behavioral traits are the product of a very complex interaction between nature and nurture, and (4) any society that is committed to accommodating and rewarding individual merit rather than demographic categories must keep its focus on the talents of individuals, not on the averages.

That said, here are just a few highlights from the technically challenging textbooks in behavioral genetics.[46] First, a vast literature on personality and behavioral traits in other species provides a theoretical foundation for the discipline. In fact, for many centuries animal breeders have successfully selected for desirable traits in many species, from horses and cattle to domesticated dogs. As every pet owner knows, our pets have distinct personalities and idiosyncrasies. Much of the early scientific research in behavioral genetics used animals, such as inbred strains of laboratory mice (*Mus musculus*), to tease out genetic differences. The "mouse men" (as they were called) were able to show in literally thousands of experiments over the course of the twentieth century that there are genetically based differences in such mouse behavioral traits as activity levels, aggression, emotionality, learning abilities, memory functions, psychomotor skills, reproductive behaviors, and sociability.

Among humans, genetically based "anomalies" have been linked to many behavioral syndromes, including disorders like schizophrenia, manic-depressive psychosis, Down syndrome, Parkinson's disease, autism, perceptual and learning disabilities, and many others.[47] But most important for our purpose are the behavior genetic studies that have shown conclusively that roughly half, overall, of the variation we observe in normal human personality traits is due to differences among the 3,150 or so genes that, according to the Human Genome Project, are involved in the development of our brains.

Literally dozens of studies have shown that between 40 and 50 percent of the total variation among us in terms of where we fit on what psychologists refer to as the "big five" personality dimensions, as measured by various psychological tests (introverted-extroverted, conscientious-disorganized, calm-nervous, agreeable-disagreeable, and open-minded or closed-minded), is due to genetically based variation (or heritability). Especially important is that similar results have been found for many of the specific traits associated with a sense of fairness. For example, one large study involving five hundred twin pairs found genetically based variances of 51 percent for "altruism," 51 percent for "empathy," and 43 percent for "nurturance"—traits that are all associated with fairness behaviors.[48]

The overall conclusion from this research, which I will develop at length later on, is that any society must be organized so as to deal with important personality differences that may be relatively intractable, especially among those "outliers" (another Malcolm Gladwell characterization) who can either enrich a society or undermine and destroy the social contract. (These "outliers" are especially dangerous when they gain political power, as can be seen in the long, depressing list of tyrants who defile our history books.) Biological variation is the rule in the natural world. Indeed, it is one of the basic factors underlying Darwin's theory, and human behavioral variations are clearly no exception. This is why utopian leveling schemes based exclusively on manipulating the cultural environment are destined to fail.

The Brain Sciences

Phineas Gage may have been the accidental inspiration—literally—for the development of the brain sciences.[49] Gage was the foreman of a railroad construction crew that was preparing a roadbed for a new rail line near Cavendish, Vermont, in late September 1848 when a gunpowder blasting charge he was setting exploded accidentally, sending a three-foot iron tamping rod with terrific force up through his jaw and out through the top of his head. The rod landed some eighty feet away. Amazingly, Gage was still alive and conscious, even coherent, as he was escorted into town, where two doctors attended to his wounds.

Though Gage's convalescence was long and difficult, partly owing to secondary infections, over time he did heal from his severe wounds. But he was not the same person. He had undergone a radical personality change. Accounts of his subsequent behavior differ, and there has been much exaggeration and hyperbole in various accounts of this famous episode over the years. But what is fairly certain is that a young man who had been intelligent, mature, organized, hardworking, and capable of leading other men (he was the foreman of his work "gang") had turned into a person who was erratic, emotionally volatile, disorganized, foul-mouthed, and had difficulty with social relationships. His former boss would not take him back, so Gage drifted in and out of various low-paying jobs. Friends and acquaintances described him as "no longer Gage."

One of many unanswered questions about the poorly documented history of this freak accident is whether Gage showed progressive recovery over time. One fragmentary report about his behavior in Chile, where he worked for several years before his untimely death in 1860 at age thirty-six (after a series of convulsions), suggested that his behavior had become more normal. We will never know for certain.

In any case, what is indisputable about this remarkable case is that it showed how much our personality and our behavior reflect what is going on inside our brain and neuroendocrine system at any moment. To be sure, what is "going on" is always a dynamic synthesis of many influences—the neurological structures that are specified in our genetic "blueprints," the ways embryonic influences and child development processes shape the phenotype, how family and cultural experiences imprint themselves on our brains, and even the dynamics of the immediate context. Nevertheless, the brain is the hub where all these influences come together. As the neurobiologist Antonio Damasio pointed out in his important 1994 book *Descartes' Error,* the Phineas Gage incident shows that our reasoning powers and emotional sensibilities are closely intertwined. Indeed, the emerging new science of "moral psychology" assigns a primary role to our underlying "moral intuitions."[50]

In the roughly 160 years since the tragic Phineas Gage accident, the brain sciences have learned a great deal about the functional organization of the brain and how its many distinct parts interact. Entire textbooks are devoted to elucidating the anatomy of the brain. There is also a large and still expanding research literature on the physiological and behavioral consequences of various kinds of brain damage (with a distressing number of recent additions as a result of the war in Iraq). The complex biochemistry and electrical properties of the brain are now more clearly understood. We even understand the complicated electrochemical "dance" involved in the firing of individual neurons, each of which may have from five hundred to twenty thousand connections to other neurons. It is also estimated that perhaps five hundred genes are entrained in shaping how each of our 100 billion or more neurons operates. Finally, we are beginning to get a fix on the brain as a complex system—how the brain performs specific functions like vision and speech. For instance, some fifty areas of the brain are associated with how we see and interpret our surroundings.[51]

As for our moral impulses, Michael Gazzaniga, one of the world's leading brain scientists, poses the following question in his 2005 book *The Ethi-*

cal Brain: "Do we have an innate moral sense as a species, and, if so, can we recognize and accept it on its own terms?"[52] His answer, as you might have guessed, is unequivocally yes. Indeed, he and various colleagues believe that our moral life is actually grounded in our emotional repertoire rather than our reasoning powers.

He refers to this as the empathy-altruism hypothesis. "From an evolutionary perspective," says Gazzaniga, "the neural structures that tie altruistic instincts to emotion may have been selected for over time because helping people immediately is beneficial. Gut instinct, or morality, is a result of processes selected for over the evolutionary process."[53] In other words, our ethical systems are adaptations—a product of both biology and culture—which have played an important part in our collective survival and reproduction over time, just as Darwin supposed (see chapter 3).

Gazzaniga's views have been seconded by the distinguished neurobiologist Donald Pfaff in his provocative book *The Neuroscience of Fair Play: Why We (Usually) Follow the Golden Rule.*[54] One of the pioneers in demonstrating how different biochemical substances act on specific areas of the brain and how they can predictably evoke behavioral responses, Pfaff notes that we now know our brains are "hardwired" (in neuro-speak) for a broad array of behavioral predispositions that are poised to be invoked, or provoked, at any time—including fear, anxiety, anger, stress, love, and empathy, among others. Pfaff thinks our ethical impulses work in much the same way. In his words,

> I believe that we are wired to behave in an ethical manner toward others, and they toward us. But with all the life-supporting functions that the brain handles from one millisecond to the next, only a few are likely to be capable of sparking an ethical response. These must be circuits crucial to our survival: the circuits that are active whenever a situation may suddenly or significantly change our current status: [such as] when a child runs out into the street in front of your car; when you are hurrying to get out of the rain and a stranger slips on the sidewalk; when you turn on the television and see an appeal for disaster relief in a part of the world you know only from a map.[55]

Pfaff adds that he and many of his colleagues do not think there is a specific ethical module or circuit in the brain; rather, they believe ethical behaviors derive from other emotional circuitry. He also theorizes that ethical behaviors may be associated with a loss, or suppression, of information in the brain that might otherwise impel us to act in a more egoistic, self-serving way.

However, recent research using magnetic resonance imaging (MRI) brain scans has suggested that both more general emotional circuitry and certain localized areas of the brain may interact when moral concerns are being addressed. In other words, a complex neural network is involved. In a series of experiments conducted at Princeton University, Joshua Greene was able to identify several areas that are activated when moral choices are confronted: one is the cleft behind the center of the forehead that has also been associated with empathy; a second area is the superior temporal sulcus just behind the ear, which is involved in gathering information about the behavior of others; a third is the posterior cingulate, and a fourth is the precuneus, both of which are activated when strong emotions are evoked.[56]

Meanwhile, another team of scientists, led by Alan Sanfey, has identified a brain area that may be associated specifically with our feelings of fairness and unfairness.[57] Experiments utilizing the so-called ultimatum games (I will talk more about these below) triggered reactions in a small area called the anterior insula, which is also associated with feelings of anger and disgust.

Social scientist James Woodward and biologist John Allman introduce their important recent paper on moral intuition this way:

> We propose that moral intuitions are part of a larger set of social intuitions that guide us through complex, highly uncertain and rapidly changing social interactions. . . . The neural substrates for moral intuition include fronto-insular, cingulate, and orbito-frontal cortices and associated subcortical structures such as the septum, basil ganglia and amygdala. Understanding the role of these structures undercuts many philosophical doctrines concerning the status of moral intuitions, but vindicates the claim that they can sometimes play a legitimate role in moral decision-making.[58]

Although we are still at an early stage in understanding the neurobiology of moral behaviors, the door to further progress is now wide open.

Evolutionary Psychology

This relatively new subdiscipline of psychology aspires to explain a wide range of psychological phenomena—from our common fear of snakes to our distinctive mating patterns, male dominance competition, and incest avoidance—in terms of natural selection and our evolutionary history as a species. Traditional psychology has been largely descriptive and is focused on identifying regularities and "mechanisms" of behavior, whereas evolutionary psychology also strives to answer *why* particular patterns of behavior exist in humankind, and what are their adaptive functions in relation to our survival and reproduction as individuals, families, and social groups. The founders of evolutionary psychology, John Tooby and Leda Cosmides, point out that this approach fulfills the vision for a natural science of psychology that Charles Darwin himself suggested in *The Origin of Species*. As Tooby and Cosmides put it, "The long-term scientific goal toward which evolutionary psychologists are working is the mapping of our universal human nature."[59]

Two research domains in evolutionary psychology are especially relevant here. One is the work of Cosmides and Tooby, and a number of their colleagues, on what they term "social exchange" (reciprocity). They note that social exchange is both pan-human and very ancient. "It is found in every documented culture past and present and is a feature of virtually every human life within each culture, taking on a multiplicity of elaborate forms such as returning favors, sharing food, reciprocal gift giving, explicit trade, and extending acts of help with the implicit expectation that they will be reciprocated."[60] Cosmides and Tooby also adopt the term social contract, which they define as any conditional or "if-then" social relationship: If you scratch my back, I'll scratch yours.

Cosmides and Tooby believe that reciprocity is an important element of human nature; there are specific neurocognitive features in the human brain (what they refer to as "mental modules") that are designed for reasoning about social exchanges and, especially, for detecting "cheaters." Over the years Cosmides and Tooby, along with a number of other researchers, have done extensive experimental research on this issue in various societies, using psychological testing tools such as the Wason selection task, and they have shown that expectations of reciprocity and cheater detection are highly specific skills that seem to involve a "dedicated system." Even three-year-olds are very good at it, and schizophrenics with impaired general reasoning abilities nevertheless are still able to detect cheaters in social exchanges. Humans are also very good at differentiating between accidental and intentional "defection" (as I noted in chapter 2).

The other research domain in evolutionary psychology that is important to a science of fairness is the work of Dennis Krebs and others on the evolution and development (ontogeny) of morality—again affirming one of Darwin's insights and building on the pioneering work of psychologists Jean Piaget, Lawrence Kohlberg, and others. Not only is morality a universal trait in the human species, but it develops in each child in distinct, well-understood stages and plays a critical role in our social relationships. As Krebs explains, "The biological function of morality is to uphold fitness-enhancing systems of cooperation by inducing members of groups to contribute their share and to resist the temptation to take more than their share. . . . Morality boils down to individuals meeting their needs and advancing their interests in cooperative ways." There is nothing immoral about pursuing your personal survival and reproductive needs, Krebs points out. Morality relates to the means you use to do so and their impact on others.[61]

Experimental and Behavioral Economics

What Eric Beinhocker, in his paradigm-changing book *The Origin of Wealth,* refers to as "traditional economics" represents a set of ideas, backed by mathematical models, that have dominated economic theory for over a century and have deeply influenced generations of economists and their students.[62] This orthodoxy is grounded in the core assumption that people will act "rationally" in pursuing whatever their interests may be, meaning that they will seek out the most efficient and effective outcomes. This model works well enough in many economic transactions, but it is insufficient. Very often people don't optimize their choices. They merely "satisfice," to use systems scientist Herbert Simon's famous term, or they exhibit what is referred to as "constrained rationality."

Equally important, orthodox economic theory focuses on procedural rationality—how best to accomplish one's "revealed preferences," whatever they may be. Traditional economic theory has had little to say about substantive rationality—the underlying motives that animate our preferences. This is one reason economics has remained a science with limited predictive power. It strives to determine what people will do if they are "efficient" in the pursuit of their personal preferences.

However, as we have lately observed, sometimes people behave in very irrational ways. Twentieth-century economist John Maynard Keynes's famous term "animal spirits" (perhaps inspired by Plato) has recently come back into vogue.[63] It was highlighted especially in a 2009 book, *Animal Spirits,* by economists George Akerlof and Robert Shiller. In the authors' words: "Much economic activity is governed by *animal spirits.* People have non-economic motives. And they are not always rational in pursuit of their economic interests."[64] Fairness is one of these noneconomic motives, they argue.

Even before the current economic crisis, however, the landscape of economic theory had begun to change. One example is the rise of modern welfare economics (which I will talk about in the next chapter). The work of Akerlof and others on the role of customs and norms in shaping economic behavior is another important example.[65] Complexity science, which is focused on the emergent dynamics of economic activity and is championed by Beinhocker, is yet another example. But especially important for our purpose here has been the emergence of behavioral economics and experimental economics—new subdisciplines that study how people actually behave in economic transactions. And one of the most powerful research tools for studying economic behavior is game theory.

"Game theory" is really a misnomer. It should perhaps be called "social interaction theory," for it is not primarily about playing games. It is about the conditions under which two or more persons will cooperate and, equally important, the contexts within which the "players" will compete and even exploit one another (conflict situations). Some are referred to as zero-sum games, because the pluses for one player in a two-person game are equal to the minuses for the other player, so that the sum nets out to zero.

In other words, game theory is really about social relationships and their consequences when people behave "rationally" (in accordance with the fundamental behavioral assumption of mainstream economics and game theory). After a slow start in the 1940s and 1950s, more or less as a theoretical sideline that deployed very narrow, constricting assumptions, game theory has blossomed into an increasingly potent tool with broad application to many economic, social, political, and even evolutionary problems in theoretical biology. (One notable real-world application was a United States government auction of rights to radio frequencies for cell phones a few years back, where the use of game theory principles netted the government a profit of $20 billion.)

Early on, game theory got a bad name in many quarters because it was focused on how to beat your opponent—zero-sum games. Indeed, the founding father of game theory, the great Hungarian American mathematician John von Neumann, was sometimes viewed as a model for the ghoulish Dr. Strangelove in the classic 1960s movie about an accidental nuclear war. (Another frequently named model was the nuclear war theorist Herman Kahn.) Von Neumann's friends assure us that he was a genial and kindly person.

One of the leading game theorists of the current generation, Ken Binmore, has pointed out that the progress of game theory was seriously retarded because the only construct that early theorists used to model cooperation was the "Prisoner's Dilemma." As Binmore observes, "A whole generation of scholars swallowed the line that the Prisoner's Dilemma embodies the essence of the problem of human cooperation. . . . On the contrary, it represents a situation in which the dice are as loaded against the emergence of cooperation as they could be. . . . Rational players don't cooperate in the Prisoner's Dilemma because the conditions necessary for rational cooperation are absent." He notes that it's rather like explaining why people drown when they are thrown into Lake Michigan with their feet encased in concrete.[66]

Over time this highly restrictive, dead-end approach to game theory began to give way as a new generation of theorists sought out better ways to model and explain how rational players can achieve stable cooperation. A cornerstone of the new approach is the Nash equilibrium—a concept developed by the Nobel prize–winning mathematician John Nash (whose mental illness and troubled life were reenacted in the movie *A Beautiful Mind*). In essence, a Nash equilibrium refers to a social relationship in which there is a stable, self-reinforcing pattern of cooperation. This can occur when all the participants are benefiting as much as they can (optimizing) and no further gains are possible for anyone by changing strategies.

Like Plato's ideal state in *The Republic* and Karl Marx's idyllic communes, a Nash equilibrium is an idealization. It assumes that everyone is perfectly rational and equally powerful, and that nobody has an incentive to cheat. Nevertheless, it does illuminate what lies at the core of any stable cooperative relationship when people are rationally pursuing their own self-interest in concert with others and doing so voluntarily. Indeed, in many of our personal relationships we are unconsciously pursuing something like a Nash equilibrium strategy.

However, this raises another question: A social relationship may be voluntary and stable, but is it fair? Some game theorists claim this is so by definition. If it entails a voluntary arrangement, it must be fair, and that is all you can ask game theory to do. I would argue, to the contrary, that game theory—like economic theory itself—is myopic about substantive fairness and remains unrealistic in one fundamental respect. In its classic form, it assumes that the players are free to choose what is in their best interest without any constraints that may adversely affect their decisions.

In the real world, an individual's "voluntary" decision may be biased by power relationships—physical, economic, or psychological coercion—and by the potential costs of *not* cooperating. For instance, many millions of workers in various countries are more or less forced to accept inadequate wages and harsh working conditions, not out of free choice but because they must take whatever jobs are available or else starve to death. So coerced cooperation may be procedurally stable but substantively unfair. As Binmore concedes, game theory "has no substantive content. . . . It isn't our business to say what people ought to like." (I will return to this crucial issue in chapters 6 and 7.)

Another major milestone in the development of a more realistic approach to game theory involved the now-famous tournament arranged by political scientist Robert Axelrod and biologist William Hamilton in the early 1980s.[67] They addressed the problem of how cooperation can get started in the first place and how it can be maintained over time—a familiar real-life problem. Using a kind of tournament format, Axelrod and

Hamilton asked a number of their colleagues to propose solutions, and the winning answer, submitted by the pioneer game theorist Anatol Rapoport, was called tit-for-tat—or reciprocity! In other words, game theorists had rediscovered the Golden Rule. Do unto others what you would like them to do for you and repay the other player in kind.

According to the tit-for-tat model, if you start out in a relationship by offering to cooperate, and your prospective partner responds cooperatively, an ongoing pattern of mutual cooperation can easily be initiated (as most of us know from experience). On the other hand, if the reciprocator responds by "defecting" (or exploiting the cooperator), the cooperator should be exploitative in the next "round" and "educate" the cheater (as many of us in fact do). Then, if the initial defector learns from the punishment and in a subsequent round shifts to cooperation, a stable pattern of cooperation can finally emerge. Thus, the tit-for-tat game can also be "robust" against exploitation, or cheating.

Among the problems with the tit-for-tat game, as various critics have pointed out, is that it requires the participants to tolerate defectors and cheaters—to turn the other cheek in hope of a change of heart—and to continue playing the game no matter what. So Martin Nowak and Carl Sigmund developed an alternative to tit-for-tat that they called "win-stay, lose-shift."[68] As the name implies, in this game a cooperator is free to leave the game altogether and play only with those who are trustworthy cooperators (which, again, more closely resembles real-life situations).

Still more realism was added to game theory with the emergence of what is called "strong reciprocity theory"—or tit-for-tat with teeth.[69] Among the limitations that continued to hamper the development of game theory was the assumption that the players were not effectively able to deter or punish defectors and cheaters. So game theory was unintentionally biased in favor of cheating. Yet in the natural world, as in every human society, punishments for transgressions are a commonplace. They have been documented even in social insects, as well as in many mammals and primates.[70]

Strong reciprocity theory rectifies this shortcoming. A clutch of theorists, including Herbert Gintis, Samuel Bowles, Ernst Fehr, Simon Gächter, Joseph Henrich, Robert Boyd, Peter Richerson, Carl Sigmund, and others, have amassed a large body of experimental evidence showing that even altruistic behaviors can be elicited in cooperative situations if there is a combination of strict reciprocity and punishment for defectors. These theorists conclude that strong reciprocity is one of the core elements of human morality and that it has played a vital role in our evolution as a species.[71]

Their conclusion is supported by experiments using a relatively new class of games, the best known being the so-called ultimatum games (again, this is a misnomer—they are really sharing and fairness games). In the typical ultimatum experiment, two players have different assignments. One player is given a quantity of money and is directed to share a portion of it at his or her discretion with the other player, subject only to the proviso that, if the recipient thinks the percentage split is too low, he or she can reject the offer and then both participants will lose out on sharing the reward.

Now, classical economic (rational self-interest) theory would predict that the sharers will offer the lowest possible amount—maybe 10 percent or less—and that the recipients will accept whatever share they can get because it is better than nothing. But this doesn't happen as a rule. Typically, sharers will offer about half of the total amount, and recipients will reject offers of 30 percent or less. In other words, both participants are operating out of a sense of fairness and, most surprising, recipients will altruistically sacrifice a reward in order to punish someone who is acting unfairly.

This outcome utterly contradicts the selfish *Homo economicus* model, though it has been replicated innumerable times, including a systematic cross-cultural study in fifteen societies by anthropologist Joseph Henrich and his colleagues. (A later study led by Henrich identified significant cross-cultural variations, however.)[72] The ultimatum research represents one of the most stunning confirmations we have that a sense of fairness is universal in the human species. Altruistic sharing backed by a threat of punishment for selfish violations is a fundamental element of human nature, coupled with the strong expectation for reciprocity from others (again, with the exception of the outliers who can undermine any stable social pattern).

Fairness, and strong reciprocity, is also relevant for what has recently become an important concept in the social sciences, namely, "social capital."[73] Although there have been a bewildering variety of definitions and interpretations of this term over the years, the common denominator is the idea that social relationships, and especially social networks, can have real value and add productivity to any purposeful activity. Like its namesake, economic capital, social capital is not a thing but a *capacity*—the wherewithal to help us achieve our goals in contexts where we can benefit from cooperating with others. This could apply to an individual looking for a job with the help of friends (and their friends), or it could involve a large organization with a tightly coordinated combination of labor. Any group is more effective, it is clear, when its members cooperate effectively. However, effective cooperation depends on trust, a sharing of effort, and strong (reliable) reciprocity. In other words, an ethic of fairness is the glue that binds social networks together and sustains social capital. (I will return to this important concept in the final chapter.)[74]

There is one other important game theory insight that I should also mention. It happens that many forms of cooperation are self-policing; they depend on the fact that all participants must give their unstinting cooperation or else the benefits cannot be realized. This was illustrated by biologists John Maynard Smith and Eörs Szathmáry in their book *The Major Transitions in Evolution,* using a metaphor from boating.[75] In a "sculling" situation, where two oarsmen are rowing a boat together and each has a set of two oars, it is possible for one oarsman to slack off (cheat) and let the other do most of the work (though in real life we are very good at detecting cheaters, as Cosmides and Tooby have shown). However, in a "rowing" situation, where each oarsman has only one of two oars mounted on opposite sides of the boat, if either oarsman slacks off the boat will go in circles and will not reach its destination. The outcome involves what is referred to as a functional interdependence. As Maynard Smith and Szathmáry conclude, this may be the most common way cooperation evolves in the natural world.

Perhaps the most balanced synthesis of the state of the art in our quest for understanding the nature and nurture of morality was provided by the psychologist cum biologist cum anthropologist Marc Hauser in his acclaimed and deeply documented 2006 book *Moral Minds: How Nature Designed Our Universal Sense of Right and Wrong.* Hauser proposes that our moral behaviors can be likened to our language abilities, where (following Noam Chomsky's theory of language) there is a "deep structure" of evolved predispositions that vary in their expression, depending on our particular language and cultural context—and personality. Taking an ecumenical stance among the competing theories of morality, Hauser makes a case for the view that our moral actions are a compound of our intuitions (and "passions"), the cultural "norms" that surround us, and our reasoning powers in specific situations. "I argue that moral judgments are mediated by an unconscious process, a hidden moral grammar that evaluates the causes and consequences of our own and others' actions. This . . . shifts the burden of evidence from a philosophy of morality to a science of morality."[76]

*

To summarize then, the science of human nature, and the subdiscipline of fairness research, are still far from having a full, and fully coherent, picture of what makes us human. It's rather like a very large jigsaw puzzle where most of the border and some of the inside sections have been filled in. The overall framework is becoming better defined, but much work still needs to be done to complete the picture.

Our evolutionary prehistory and the emergence of complex civilizations over the past fifty thousand years, as well as our turbulent recent history as a species, all confirm that humankind represents a bundle of paradoxes. Plato and Aristotle well understood the nature of these paradoxes, even if their models were intuitive and rather crude. The science of economics has established beyond dispute that we are very often absorbed in our own needs and wants. We are undeniably self-serving, as is the case for every other living organism. It's a biological imperative. Many of us are also highly competitive and strive for status and power. Moreover, some of our evolved "appetites," which are after all designed to serve our survival and reproductive needs, may misfire and become self-destructive. The seven deadly sins are not just a Sunday sermon topic.

On the other hand, we are deeply embedded in complex social environments where service to others and working with others to satisfy our common needs are totally consistent with our self-interest, another insight that traces back through Adam Smith to Plato and Aristotle. We also care deeply about social acceptance and approval, because the survival and reproductive success of our ancestors depended on it for millions of years. Also, we display extreme sensitivity to being treated unfairly by others and are, as a rule, receptive to others' claims for fairness.

Yet, as I noted earlier, our sense of fairness seems to have a sharply defined psychological boundary. We readily extend help and may even display unconstrained altruism toward others we identify with as members of our own "tribe." But we can be indifferent to the sufferings of outsiders and may be quite willing to inflict harm on them if it suits our interests. Behavioral economists call it "parochial altruism."

A further paradox is that all of these deeply rooted appetites and emotions are highly malleable. Our destiny as individuals and as superorganisms is shaped both by nature and by nurture. We have many evolved predispositions, but we are able to exert some control over their expression, and this also applies to the pursuit of fairness and social justice. Our sense of fairness is embedded in our human nature, and we are (most of us) predisposed to act in ways that conform to the Golden Rule. Along with other aspects of our morality, our fairness motives may be based either on altruism—uncompensated acts of charity—or on enlightened self-interest, including a concern for social approval, expectations of reciprocity, or both. But the specific content of our actions is also shaped by a combination of social norms, the immediate context, and the seductive lure of self-interest. We are highly vulnerable to the calculus—often subconscious—of what serves our personal agendas. All too often we find rationalizations to justify rejecting other people's fairness claims.

The science of human nature has also confirmed the necessity for taking into account the intractable fact that biologically based individual differences also shape our personalities and behavior, for better or worse. This is most obvious when we are dealing with the extremes—the ruthless egoists that almost everyone in the business world has had to deal with (often in court), as well as the sociopaths and psychopaths for whom fairness is an incomprehensible concept. Conversely, our very real differences in terms of talent, ability, and personal accomplishments are very pertinent to how we define fairness. When we undertake the task of formulating a new bio-social contract later on, we will need to take account of these individual differences.

Finally, I must also draw attention to the fact that there is a huge and profoundly important omission from the jigsaw puzzle of human nature, as it has been characterized (as a rule) by our social scientists. Often they overlook the most fundamental questions of all: What is the underlying purpose of human nature? What are the sources of our interests—the roots of our "revealed preferences"? Indeed, many economists have treated our basic needs as being none of their business. In a tradition that harks back to the classical economists and the philosopher David Hume, they claim that no such rendering of our basic needs is scientifically possible, or proper. It trespasses into the forbidden (normative) realm of specifying what we ought to do. Our "tastes" differ, we are told.

In the next chapter I will address this concern and will argue that in the end we cannot understand the full dimensionality of the collective survival enterprise—much less the biological basis of fairness—without specifying in detail the array of survival and reproductive needs that have preoccupied our ancestors for millions of years and that still preoccupy most of us today. We need to get a better fix on the concept of "basic needs."

5 * *Human Nature and Our Basic Needs*

> I do not think we have adequately determined the nature and number of the appetites, and until this is accomplished the inquiry [about justice] will always be confused.
>
> SOCRATES, IN PLATO'S *Republic*

The fundamental problem for every living organism is survival (and reproduction), and humans are no exception. To repeat, whatever may be our perceptions, aspirations, or illusions, biological survival remains an inescapable imperative; life is at bottom a contingent survival enterprise.

This is the underlying purpose and vocation of human nature. Though we do not often dwell on this fact, we live every day in the shadow of death and all too often bear witness to the fragility of life. Indeed, many of the things our species has been doing collectively are unsustainable over the long run and are jeopardizing our future survival as a species. On the other hand, consider our past achievements. And look also at what our more remote ancestors accomplished. We are the latest links in an unbroken chain of life that is more than 3 billion years old. As the prominent twentieth-century biologist Julian Huxley so eloquently expressed it, we are "evolution become conscious of itself."[1]

Nonetheless, life remains a formidable challenge, and "earning a living" is—quite literally—the central organizing principle for every human society. Most of what occupies our lives as a species is related, either directly or indirectly, to the ongoing survival enterprise, even though it may be the furthest thing from our (conscious) minds as we go about our daily activities. Biologists call it adaptation and use the term fitness to characterize how well (or poorly) any organism, humans included, is coping with the problem.

The pioneering twentieth-century Russian American biologist Theodosius Dobzhansky was fond of characterizing the evolutionary process as "a grand experiment in adaptation."[2] And Julian Huxley, in his landmark 1942 volume *Evolution: The Modern Synthesis,* defined adaptation as "nothing else than arrangements subserving specialized functions, adjusted to the needs and the mode of life of the species or type. . . . Adaptation cannot but be universal among organisms, and every organism cannot be other than a bundle of adaptations, more or less detailed and efficient, co-ordinated in greater or lesser degree."[3] In other words, adaptations have a purpose: they are "designs for survival," in biologist George Williams's characterization.[4]

The basic theoretical challenge, then, is how to measure biological adaptation and fitness in human societies. What criteria do we use? And how can we differentiate between better and worse forms of adaptation? Unfortunately, "traditional" economic theory (in Beinhocker's term) cannot be of much help to us. Orthodox economics has systematically obscured, and sometimes even denied, this biological reality behind a smokescreen of euphemisms and technical abstractions—utilities, tastes and preferences, demand and supply, externalities, and even ill-defined measures of well-being and happiness.

These metrics are at best only indirectly related to our biological survival needs and sometimes may even mislead us by calling things economic "goods" that are biologically bad for us. Thus many economists are pleased (at least professionally) if as a society we overindulge in unhealthy foods. This represents an increase in the "demand" for food products, which genuflects to the holy grail of economic growth even though it may lead to obesity and produce an epidemic of diabetes (some 10 percent of all Americans and one-third of our children, according to recent health statistics).[5] And if our collective obesity creates an additional demand for health services, this too adds to our gross domestic product (GDP), an unalloyed good if you follow the logic of a growth-oriented economics profession.

Only now are the social sciences emerging from more than a century of "value relativism," as it was called, in which all needs, wants, and preferences were assumed to be culturally defined and therefore infinitely pliable—a viewpoint inspired by the worldview of the utilitarian philosophers and, before them, the Sophists and Epicureans (as I noted earlier). Economist John C. Harsanyi's principle of "preference autonomy" (preference utilitarianism) epitomizes this posture: "In deciding what is good and what is bad for a given individual, the ultimate criterion can only be his own wants and his own preferences."[6] The Austrian economist Friedrich Hayek—much revered in conservative circles—called it "methodological individualism." He characterized economic actors as being like empty vessels that respond mechanistically to whatever culturally defined preferences are poured into them. In fact, many twentieth-century social theorists denied the very concept of basic needs. Andrew Heywood summarized the argument as follows:

> Needs are notoriously difficult to define. Conservative and sometimes liberal thinkers have tended to criticise the concept of "needs" on the ground that it is an abstract and almost metaphysical category, divorced from the desires and behavior of actual people. . . . It is also pointed out that if needs exist they are in fact conditioned by the historical, social and cultural context within which they arise. If this is true, the notion of universal "human" needs, as with the idea of universal "human" rights, is simply nonsense.[7]

Heywood was not alone. Sociologist Angus Campbell, for instance, noted "the obvious fact" that "individual needs differ greatly from one person to another and that what will satisfy one [person] will be totally unsatisfactory to the other."[8] Likewise, sociologist Erik Allardt asserted that "a level of need satisfaction defined once and for all has hardly any specific meaning. . . . To a large extent, needs are both created by society and culturally defined."[9] Sociologist Geoffrey Rist was even more dogmatic: "Needs are constructed by the social structure and have no objective content."[10] And economists William Baumol and Alan Blinder in their introductory textbook tell us that "the fact that the concept of poverty is culturally, not physiologically, determined suggests that it must be a relative concept. . . . The basic problem with the absolute poverty concept is that it is arbitrary."[11]

In short, all preferences are created equal. It does not matter what you may wish to do, so long as you pursue it "rationally"—meaning as efficiently and consistently as possible. Of course, no academic discipline is ever monolithic, and economics is certainly no exception. Among the various side branches of this diverse profession there is one—tracing its roots back to the early classical economists—that is known as welfare economics. As the name implies, welfare economics is concerned with the end results of economic activity—how it affects the "general welfare." Much of the early work in welfare economics ducked the central issue, however. Following the lead of the utilitarian philosophers, early theorists like John Stuart Mill, Arthur Pigou, and others defined social welfare as, simply, the sum of everyone's individual "utilities," or satisfactions, regardless of how these might be distributed.

More recent theory has focused on how well an economy achieves "efficiency" in the allocation of goods and services, albeit in a very narrow sense. Thus, an economy is said to be "Pareto efficient" if no individuals can be made better off without making somebody worse off (as measured in terms of income, mainly). So, if the top 20 percent of the population holds 84 percent of the national wealth and the bottom 80 percent holds 16 percent (as is the case in America today), that's all right so long as it's Pareto efficient.[12]

In fairness, economic theorists have also developed a "social welfare function" that weights each individual's "marginal utility," so that an extra dollar of income for a billionaire counts for less than an extra dollar for someone earning the minimum wage. Such qualifiers may be tractable mathematically, but they have no practical value unless you can actually measure and compare those marginal utilities—a complex and highly subjective enterprise that many theoretically inclined economists have avoided.

One important exception is the work on equity theory in the social sciences, which is concerned with distributive justice in a narrow sense—such as how to achieve an "envy-free" division of benefits and costs in a particular context, although a variety of other principles have also been utilized. As Peyton Young, one of the leading theorists in this discipline, explained in his 1994 book *Equity*, the focus of equity theory is not on what constitutes a just social order but on such concrete social issues as property disputes, admission to law school, or charges for public services. There are, it turns out, many complexities associated with such decisions, and various tools have been developed for dealing with them systematically.[13] (I will revisit equity theory in chapter 7.)

Another notable exception to the value-free posture in economics is the work of the Nobel laureate Daniel Kahneman and his colleagues on measuring a person's overall life satisfaction, or what they call "experienced utility."[14] There is also a body of work on the concept of "well-being," where various efforts have been made to break the concept down into its component elements, or to identify global surrogate measures. Still another divergence from conventional economics has been the recent surge of interest and research on "happiness," in both economics and psychology. The challenge is to develop a measure of happiness that is independent of income, or of one's "revealed preferences" in the marketplace. Happiness researchers Bruno Frey and Alois Stutzer claim that happiness is "one of the most important issues in life—if not *the* most important issue."[15] (I'm reminded of the cynical old saying, "What good is happiness? You can't buy money with it.")

Although happiness is certainly a worthy subject, I believe that, like psychologist Abraham Maslow's well-known concept of "self-actualization," it is analogous to the top layer of a cake and does not reliably reflect what the layers underneath are made of.[16] Happiness surveys may well prove to be a useful indirect measure of basic needs satisfaction, but the argument here is that these needs should—and can—be measured directly.

Overshadowing all of these efforts, however, is the prolific and important work of Nobel Prize–winning Indian economist Amartya Sen and his colleagues.[17] In a series of writings that date back to the 1970s, Sen has mounted a major assault on the utilitarian, subjectivist model of well-being. Paralleling some of the arguments of philosopher John Rawls, Sen has challenged the adequacy of the various "psychological" formulations of welfare that rely on desires, tastes, subjective utilities, or what have you.

Sen accuses mainstream economics of circularity, vacuity, gross oversimplification, and the use of psychological premises that are without foundation. Noting, for example, that "sympathy" and concern for others can also affect a person's welfare, or that individual welfare functions can be interdependent (as highlighted in game theory), or that social commitments may affect our behavior, Sen argues that a narrow, materialistic concept of "self-interest" is not a sufficient definition of behavioral motivation, much less well-being.

Furthermore, Sen points out, being consistent in making choices is a pretty weak definition of rationality. In a famous passage from his Herbert Spencer Lecture at Oxford University in 1976, titled "Rational Fools," Sen concluded: "The *purely* economic man is indeed close to being a social moron. Economic theory has been much preoccupied with this rational fool decked in the glory of his *one* all-purpose preference ordering. To make room for the different concepts related to his behaviour, we need a more elaborate structure."[18]

Sen does not try to define what ends any individual should pursue but rather directs our attention to the *means* necessary for setting and pursuing our personal goals. He focuses on the "capabilities to function"—the nutritional benefits of food versus food per se. Sen describes it as "a particular approach to well-being and advantage in terms of a person's ability to do valuable acts or reach valuable states of being." In the current political jargon, Sen's focus is on "empowerment" rather than a person's subjective sense of satisfaction, which, as Sen notes, may or may not be concordant. Sen tells us the functionings that are relevant for well-being can vary from elementary ones like escaping mortality, morbidity, or hunger to more complex and subtle conditions such as achieving self-respect or enjoying social interactions. However, Sen demurs from proposing "just one list of functionings."[19]

Sen also addresses the issue of poverty and basic needs in his framework. He speaks of a subset of capabilities that he calls "basic capabilities," and he defines these as "the ability to satisfy certain crucially important functionings up to certain minimally adequate levels." Noting the growing literature in the public policy realm in recent years on the concept of basic needs (see below), Sen argues that the basic capabilities approach is compatible with a basic needs approach and can greatly improve on the conventional use of income as a stand-in measure of poverty.[20]

Thus Sen clearly recognizes the concept of basic needs. Indeed, he and various colleagues have been much concerned about such pressing real-world problems as hunger and global poverty. And yet Sen's paradigm does not provide any explicit theoretical basis for his "basic capabilities." Like so many other treatments of the concept of basic needs, Sen's paradigm is at once intuitively obvious and theoretically murky.

In short, what is missing in Sen's theoretical work is a way of grounding the concept of capabilities (requisites) that is both independent and directly measurable. So far as I know, Sen has declined to elaborate on his concepts in more specific detail, so they remain elusive as analytical tools for real-world situations. Sen leaves that task to others (though more recently he has been promoting work on basic needs).

How, then, can we apply and test Sen's concepts? As various critics have argued, what is required is a "substantive list" of the elements needed to sustain life and make it valuable. Sociologist Thomas Scanlon called for an "objective index" of well-being that can pass two tests: adequacy and practicality.[21]

As it happens, there have been various efforts in the public policy field to measure the quality of life more objectively, dating back at least to the emergence of the "social indicators" movement in the 1960s. While the origins of this movement could perhaps be traced to sociologist William F. Ogburn's *Social Trends* in 1929, contemporary researchers generally identify Raymond Bauer's *Social Indicators* in 1966 as the catalyst for the more recent and sustained efforts in this area.[22] After the publication of Bauer's pathbreaking book, social indicators research enjoyed a brief period of rapid and well-funded growth.

Much of the impetus for the social indicators movement arose out of a reaction against our heavy dependence on economic indicators as the criteria for societal progress or well-being (especially the GDP and per capita income). The goal of the social indicators "idealists," as they were sometimes pejoratively labeled, was to develop a broad definition of the "general welfare" that was independent of economic growth and also accounted for what economists refer to as "externalities." Perhaps the most frequently quoted statement of this energizing vision can be found in *Toward a Social Report* in 1969, a widely influential publication sponsored by the (then) U.S. Department of Health, Education, and Welfare and written principally by economist Mancur Olson: "A social indicator may be defined to be a statistic of direct normative interest which facilitates concise, comprehensive and balanced judgments about the condition of major aspects of a society. It is in all cases a direct measure of welfare and is subject to the interpretation that, if it changes in the 'right' direction, while other things remain equal, things have gotten better or people are 'better off.'"[23]

Just as economists have been able to develop a theory of economic life that provided a framework for aggregating data on the production and consumption of goods and services, so the social indicators proponents aspired to develop a coherent system of objective measures of well-being. Unfortunately, none of these efforts were rigorously grounded theoretically. All of them rested on intuitive (though often compelling) pragmatic criteria. Although there was considerable overlap among the various attempts to formulate a shopping list of basic human needs, there were also significant differences among them.[24] Because these efforts had no compelling theoretical justification, they were vulnerable to being attacked or dismissed by the many theorists who had a vested interest in value relativism and who could claim that well-being is necessarily a personal and subjective affair.

Len Doyal and Ian Gough, in their important 1991 book on basic needs, conclude: "The movement for social indicators and human development appears to have run into the sand. . . . The decline and fall of the social indicators/human development movements was due first and foremost to the lack of a unifying conceptual framework."[25] As Sen and a coauthor, Martha Nussbaum, point out in one of their essays, "The search for a universally applicable account of the quality of human life has, on its side, the promise of greater power to stand up for the lives of those whom tradition [read economic and political forces] has oppressed or marginalized. But it faces the epistemological difficulty of grounding such an account in an adequate way, saying where the norms come from and how they can be known to be the best."[26]

Doyal and Gough concur: "The earlier theoretical innovations . . . all suffer from one overriding defect. None of them demonstrates the universality of their theory, nor, on the other side of the same coin, tackles the deeper philosophical questions raised by relativism."[27] In sum, the search for a satisfactory metric for well-being and the quality of life has been severely hampered by the lack of a theoretical foundation.

Before proceeding, I should note one other important use of the concept of basic needs. Even though it has been regularly debunked by social scientists, the concept of basic needs has nonetheless played an important political role in the development of the welfare state in Western societies over the past century. Beginning in 1883, when Chancellor Otto von Bismarck established the first "social insurance" program as a way to help unify the then new German nation-state, appeals to basic needs have often figured in the enactment of various social welfare programs in Western countries. These programs include workers' compensation, public assistance, social security, health insurance, and the minimum wage, among others.

The concept of basic needs was also an explicit part of the New Deal philosophy of Franklin Roosevelt. As FDR put it in a speech in 1931, in the depths of the Great Depression, "One of the duties of the State is that of caring for those of its citizens who find themselves victims of such adverse circumstances as makes them unable to obtain even the necessities of mere existence without the aid of others. That responsibility is recognized by every civilized nation."[28]

Thus it seems paradoxical, to say the least, that the concept of basic needs has been regularly invoked in connection with social policy and regularly rejected in social theory. But if humans are not exempted from the survival problem, there is no justification for denying the reality of our basic survival needs. This is not, as the economics profession would have it, a normative issue. It is an empirical issue. We are born with an array of biological needs and built-in "oughts" that motivate and organize our behavior. One cannot fully explain, much less predict, human behavior without reference to these biologically based preferences.

To put the argument in its most general form, humankind is subject to a conditional "if-then" imperative: If we want to survive, we must actively pursue a set of specific survival-related preferences (our basic needs), or else there will be predictable (harmful) consequences. If economists can successfully penetrate the complexities and ambiguities of market prices to get to the bedrock of "real costs," it should also be possible to get to the bedrock of "real benefits."

This challenge is addressed head-on in the Survival Indicators paradigm—so named because it is designed to measure directly the degree of satisfaction of our basic survival needs. The Survival Indicators project has its roots in some research done in the 1970s on the relation between income and basic needs satisfaction for welfare recipients in the state of California. An initial attempt to develop a Survival Indicators framework and to formulate a master indicator of adaptation called a "Population Health Index" was presented in my 1983 book *The Synergism Hypothesis: A Theory of Progressive Evolution.*[29] Additional development work on this framework since then is discussed in detail in my 2005 book *Holistic Darwinism.*[30] So here I will be relatively brief.

In the Survival Indicators framework, the term basic need is used in a strict biological-adaptive sense as *a requisite for the continued functioning of an organism in a given environmental context; that is, denying this need would significantly reduce the organism's ability to carry on productive activities, reduce the probability of its continued survival and successful reproduction, or both.* So defined, basic needs are not unique to humans; the term applies to all living things. Moreover, "need" connotes a requisite where significant "harm" (that word again) will occur if it is lacking or absent. Moreover, the nature of this harm is specified in strictly biological rather than moral terms—

that is, in terms of the "normal functioning" and "productive activities" required to be able to meet our ongoing survival needs.

Several brief comments are in order with regard to this definition. One is that the concept of basic needs is not interpreted here in a narrow, physiological sense. It is not just about food, water, and shelter (as we shall see). Like other researchers in the social indicators field, I recognize that, by its very nature, the human survival enterprise is a social activity that entails cognitive-psychological needs and a need for social relationships of various kinds. But these are not viewed here as ends in themselves. As I noted earlier, we are fellow participants in a "collective survival enterprise," and many of our needs are satisfied through socially organized activities and socially defined tasks. Equally important, the Survival Indicators paradigm recognizes that basic needs have a life cycle—a trajectory that includes growth and development, reproduction, child nurturance, and aging. The temporal dimension of the survival enterprise, often overlooked in other paradigms, is reflected in several of the basic needs domains described below.

A second point is that the Survival Indicators paradigm involves a highly nuanced concept of basic needs. In particular, I distinguish between *primary needs, instrumental needs, perceived needs, dependencies,* and *wants* (or tastes and preferences). Basic needs encompass only the first two of these categories (primary and instrumental needs). Primary needs are irreducible and nonsubstitutable. One cannot substitute food for water, or sex for sleep (well, not for long). Primary needs coincide with the broad functional requirements for adaptation.

Instrumental needs, on the other hand, are the derived adaptive *means.* For instance, we have a physiological need for a defined quantity of uncontaminated freshwater (a "primary need"), as well as an "instrumental need" both for a source of freshwater *and* for appropriate water technologies—what Sen would call a "capability"—to obtain the water and satisfy this primary need. Instrumental needs support our primary needs and may be satisfied by various functional equivalents (e.g., eating beef versus chicken or eggs as a protein source). Instrumental needs may also vary widely depending on the particular adaptive context.

It is also important to distinguish between needs and our so-called drives, or internal sources of motivation. Needs are functional requirements; drives are psychological mechanisms that we may perceive as needs. Human sexuality involves a drive that we sometimes colloquially call a need, but in reality it is an evolved instrumentality for serving our primary reproductive need. The distinction between the two concepts (need versus drive) is clearly evident, in different ways, both in the practice of birth control and in artificial insemination, where sex and reproduction are decoupled. By the same token, a person may eat either more or less than is nutritionally necessary in response to the promptings of hunger.

Accordingly, in this framework our various motivational states are distinct from basic needs and are categorized under *perceived needs, dependencies,* and *wants.* The litmus test for a *primary need,* according to this formulation, has nothing to do with whether the need is reflected in our psychological motivations or "preferences" (although most are). Nor does it matter that our primary needs vary—as they do in systematic ways that are more or less well understood. More important is that they are directly linked to the potential for suffering "harm" in the strict biological/survival sense.

A special word is in order here regarding the role of income as an instrument for basic needs satisfaction. Income is often used as a surrogate social indicator, but there are many problems associated with this approach. Sen, for one, argues strongly against using an income-based measure of well-being. On the other hand, income is also a necessary prerequisite (a means) for meeting basic needs in a great many human societies, as numerous social indicators theorists have recognized. It is therefore highly relevant as an *instrumental need,* even though it is inadequate as a summary measure of primary needs satisfaction, much less of well-being or happiness. (A number of other preliminary points about the Survival Indicators framework are included in an extended endnote for this chapter.)[31]

The fourteen primary needs "domains" (so called because several of them have more than one element, or aspect) included in the Survival Indicators framework are *thermoregulation, waste elimination, nutrition, water, mobility, sleep, respiration, physical safety, physical health, mental health, communications (information), social relationships, reproduction, and nurturance of offspring.* (The order of their presentation below is somewhat arbitrary; all these needs are viewed as equally important in terms of their relation to adaptation and fitness.)

I believe these fourteen domains represent an irreducible and indispensable requirement for biological adaptation/fitness in the human species. I should also emphasize that these fourteen categories are not ad hoc or arbitrary, but neither do they have the status of Mosaic law. They were initially formulated more than a decade ago, and in retrospect they still appear to be valid. (Only one item—respiration—has been added to the list more recently). Nevertheless, the Survival Indicators framework remains open to challenge and revision at any time if more, or fewer, or different categories can be justified.

The underlying thesis, then, is that successful adaptation and fitness in humankind are a direct consequence of meeting these fourteen primary needs. Conversely, failure to meet any one of these needs will result in varying degrees of "harm"—a reduction in the ability to engage in "normal functioning" and the pursuit of "productive activities." However, these criteria do not fully determine (or "predict") ultimate reproductive fitness. The satisfaction of our basic needs is necessary but not sufficient. Other things being equal, however, the chances of future survival and reproductive success should be much greater for those whose basic needs are fully satisfied.

Some of these fourteen primary needs domains may seem self-evident. Many of them can be found on other lists of basic needs. (We are not, after all, venturing into unexplored territory.) Other needs may appear puzzling or vague (or controversial) and may call for some elaboration. In actuality, there are complications with every one of these needs, some of which are viewed very differently from more conventional treatments. Accordingly, I will discuss each need in turn, although here it is possible to provide only a brief explanation.

Thermoregulation

Maintaining our body temperature within a very narrow range is at once a starting point and a prime example of the concept of a primary need. We often take this biological imperative for granted or respond almost reflexively to various assaults on our internal thermostats. Yet thermoregulation is a critical need—even though its demands on our time and resources are obviously context-dependent. Thermoregulation presents a very different problem in the tropics or the midday heat of Death Valley than it does in the Arctic, or even between summer and winter or night and day in many locations. The thermoregulation problem also differs significantly for someone who is inactive or asleep versus a runner or cross-country skier in full stride. By the same token, there are well-documented physiological differences between individuals in their susceptibility to cold and heat and their comfort levels. Nonetheless, this ongoing, inescapable need (even at times when we are for the moment relatively comfortable) illustrates that our primary needs are objectively important and play a direct role in shaping our cultural patterns (and technologies) as well as our economic choices (and satisfactions). A portion of our daily activities typically involves deploying instrumental means for thermoregulation. Our material (technological) adaptations can range from suitable clothing and shelter to the use of personal fans, shade trees, warm bedding, fossil-fuel heating systems, air-conditioning, and so on. Relevant (nonmaterial) cultural practices can range from huddling (sometimes even with animals) to sharing "bundling beds" (very popular in colonial times), fire building, going for a swim, or taking a midday siesta. Sometimes thermoregulation may involve taking off clothing or turning off heaters. And whenever this need is not satisfied, for whatever reason, the consequences can range from a mild disruption of one's normal routine to a high-priority search for relief, health-threatening heat strokes, frostbite, or even death from exposure.

Waste Elimination

This is another primary need that we often take for granted, even though it involves a nontrivial part of our economic activity. (Waste disposal here refers both to bodily wastes and to the various liquid, solid, and gaseous wastes produced by our personal, social, and economic activities.) Like thermoregulation, this need is conspicuous for its potential to affect other primary needs, ranging from physical health to the availability of uncontaminated air and water, sometimes sleep, and even thermoregulation (for anyone who has ever had to visit an outdoor privy in the middle of a cold night). Here again we can see that a primary need has inspired a large number of instrumental needs, especially in complex societies. And here also the problem of measurement refers to the ability to provide adequate sanitation at the individual level (by no means a given in many societies even today), as well as an appropriate infrastructure of services and technologies—from sewer systems to garbage trucks to smokestack scrubbers. Anyone who has experienced a prolonged garbage strike can testify that waste removal is an indispensable basic need in a complex society. And this example is trivial compared with the health consequences of environmental pollution. In other words, complex industrial systems have greatly expanded the scope of our primary need for waste elimination, and with it the range of instrumental means needed to satisfy it.

Nutrition

It is a safe bet that nutrition is included on virtually every social indicators shopping list. Yet, as I noted earlier, even an obvious primary need like nutrition has many components, many variables, and even perhaps some remaining unknowns. Appropriate quantities of calories are not enough to satisfy this need, no matter how many are available to us—a point underscored recently when it was reported that the Chinese diet is seriously deficient in iodine, resulting in a very high incidence of mental retardation in that country. And this is only the latest example in an age-old litany of nutritional ignorance and its maladaptive consequences. (Remember scurvy?) Indeed, malnutrition of one kind or another remains a serious problem in many parts of the world, despite our much better understanding today of what constitutes an adequate diet. Conversely, it is possible to consume too much of a good thing—sugars, fats, and overdoses of certain vitamins being especially notable problems in some developed societies. Not only does a simple term like "nutrition" mask the complexities involved in providing for this primary need, but it does not even begin to account for the vast human enterprise and the enormous range of absolutely essential instrumental needs that our nutritional needs also depend on. The list includes, among other things, fertile soil, a suitable climate, water, irrigation systems, fertilizers, seeds, tools in great profusion, farm machinery of great complexity, pesticides, animal husbandry, processing and packaging industries, and transportation and distribution systems, plus exogenous energy inputs of various kinds to power farm equipment, move water, fuel transportation systems, make fertilizers, process foods, and, not least, cook the many foods that would be toxic or infectious if eaten raw. With the exception, perhaps, of the few remaining hunter-gatherers, pastoralists, and subsistence horticulturalists (and even they depend on primitive technologies), the rest of the world's economies depend on a formidable array of food production technologies, and this says nothing about the enormous quantity of information and human skill that is also involved. If there is food on your table (or in your local restaurant) tonight, it is only because a vast human "food chain" performed its job with only minor glitches. In this light, it is a bit fatuous to claim, as some theorists do, that this basic need is not an important consideration in a developed economy. It is an adaptive modality (in the strict sense) that vast numbers of people participate in worldwide, and it is a relentless daily challenge everywhere (as we are reminded when there is a drought, a freeze, a flood, a hurricane, a famine, a trucking strike, or contamination of the food supply). Whether this vast system can be sustained, much less augmented, to deal with long-term population growth remains to be seen.

Water

Any list of social indicators that lumps food and water together can be charged with being a bit cavalier about the distinct challenges associated with providing freshwater resources and the multifaceted use that humankind makes of freshwater (and sometimes also salt water). Our need for uncontaminated drinking water is incessant and often urgent, and so are the water needs of the plants and animals that sustain us. These needs are obvious, of course. But water also serves other primary and instrumental needs, particularly those related to personal hygiene, food processing and preparation, waste disposal, and firefighting, not to mention a plethora of important manufacturing processes. Accordingly, the instrumental problems related to water acquisition, storage, transportation, and pollution have played an important role in the evolution of human societies ever since the dawn of civilization, and probably earlier. Our need for water has been responsible for many instrumental technologies over the centuries, including catchment basins, wells, irrigation systems, water containers of various kinds, pumps, hoses, viaducts, baths, sewer systems, dams, desalinization technologies, the soft-drink industry, and more. And again, the problem of meeting this primary need every day is an ongoing challenge. Potable freshwater resources are currently diminishing worldwide, and some experts in this area consider freshwater supplies to be the most critical limiting factor for the continued growth of human populations.

Mobility

The ability to purposefully change locations is a universal primary need in virtually all species of animals. (To a limited extent, this is also technically true of plants with regard to the dispersing of pollen and seeds and the spreading of roots and sometimes of other propagules.) Moreover, mobility is at once a universal primary need in humankind and an instrumentality that is critical to our ability to satisfy other primary needs. It is a need that can easily be taken for granted (it figures in relatively few lists of social indicators), yet it should not be overlooked. We are reminded of this when we observe people who are immobilized—those who are hampered by a variety of congenital defects, or the victims of land mines, or paraplegics who have suffered war injuries or debilitating accidents, or handicapped and elderly persons with various maladies. Equally important, mobility is a primary need whose scope has greatly expanded in complex modern societies. Our panoply of transportation technologies—horses, bicycles, automobiles, mass transit systems, wheelchairs, maybe even sidewalks—are instrumental needs in the strict sense; many of us would be unable to meet our basic needs without daily access to these technologies. And once again this is driven home to us when the particular technology we may depend on fails us for whatever reason—a transportation strike, a snowstorm, a fuel shortage, a malfunctioning automobile, a fare increase, and so on. Although substitutions are frequently available, or changes in lifestyle can be made to mitigate the need, the need for mobility itself is, in fact, an incessant human preoccupation. We never "solve" it; we are compelled to cope with it every day.

Sleep

If mobility is often overlooked as a primary need, sleep may seem even more problematic—except that human societies, and human behaviors, are significantly shaped by this need. Almost all of us spend approximately one-third of our lives sleeping, and a great deal of economic activity (and inactivity) worldwide is oriented to providing for this need, from work and production schedules to beds and bedding, bedrooms and sleepwear, alarm clocks and sleeping pills, and of course hotels, hostels, campgrounds, and sleeping accommodations on planes, trains, and ships. Indeed, even wars are fought with due respect for this basic need. Nor is sleep a need whose satisfaction can be taken for granted. A host of factors may interfere with our sleep requirements: work or academic pressures, family stresses, illness, insomnia, jet lag, sleep apnea (snoring), noisy neighbors, night shifts, late-night television, and so on. The consequences can range from mild fatigue and loss of efficiency to life-threatening effects on personal health or even fatal accidents. And at the risk of belaboring the obvious, it is a biological/adaptive need that can never be "solved." In fact, a recent report on sleep deprivation in the United States characterized the problem as "epidemic" in scope. Two national polls and a congressional study have concurred that (as of 1998) an estimated 70 million Americans were having trouble getting adequate sleep. More than 40 million of these suffered from potentially serious sleep apnea (often exacerbated by obesity and a lack of exercise), and another 20 to 30 million Americans were afflicted by any one of eighty other recognized sleep syndromes. The estimated loss in national productivity, in dollar terms, may be as high as $50 billion to $100 billion. Not surprisingly, there are now at least three thousand sleep clinics in the United States. It is a growth industry.[32]

Respiration

This primary need was added to our original list after it was pointed out to us that adequate respiration cannot, after all, be taken for granted and that instrumental means are often required to satisfy it, especially in modern societies. Respiration, and the provision of an adequate supply of clean, oxygenated air, can be a serious problem at high altitudes, in enclosed spaces (mines, submersibles, and modern habitations like high-rise buildings), during a fire, or when swimming. Suffocation, for various reasons, is a significant cause of accidental deaths each year, and air pollution has become a serious health threat in various localities. Moreover, there is no known substitute for respiration (in humans, at least).

Physical Safety

Avoidance of physical injury or death is included on most social indicators lists; its relationship to any definition of basic needs, well-being, or happiness is self-evident. It is also an example of a need that is greatly affected both by internal, personal influences (and behaviors) and by a host of external influences. Personal factors include such things as fatigue, alcohol or drug use, forgetfulness, distractions, errors in judgment, self-inflicted accidents, failure to use available information (such as not reading instructions), and not least, a broad range of deliberate behavioral choices, from engaging in risky sports to flying military jets or committing crimes. External influences on personal safety are equally wide-ranging. To name a few: faulty systems, bad weather, earthquakes, fires, others' mistakes, personal conflicts, feuds, deliberate acts of criminal violence, government violence, terrorism, and wars. Clearly, physical safety is an ongoing primary need in any society, and the instrumental means for mitigating threats to it are almost endless. They range from personal factors like lifestyle choices, proper education, and corrective lenses to subtleties like handrails and properly lighted stairs, automobile seat belts, effective police protection, low unemployment rates, a strong military capability, and even tranquil foreign relations. Although we often do not connect such far-reaching aspects of our economic and political systems with this primary need, they are nevertheless highly relevant.

Physical Health

This primary need is so obvious that many theorists, including a number of workers in the social indicators field, consider it virtually equivalent to a definition of basic needs satisfaction (and even well-being), or at least a good surrogate indicator. To be sure, physical health is a state that can be affected by many other variables: genetic defects, prenatal assaults, postnatal diet, immunizations, housing conditions, public health measures, health education, lifestyle, family and social relationships, job (or jobless) stresses, and much more. It is certainly highly pertinent to biological adaptation and even to well-being. (Indeed, health researcher Richard Wilkinson has documented extensively the role of chronic stress in health/disease patterns and has linked it to the many economic, social, and political conditions that can contribute to personal stress.) On the other hand, physical health is also a highly labile concept; much depends on how "health" is defined and measured. Some theorists define it very broadly, in the spirit of the well-known World Health Organization definition: "a state of complete physical, mental and social well-being." Consequently, these theorists tend to be expansive in their choice of health indicators. I prefer to define physical health much more narrowly (and conventionally) as the absence of "inborn errors of metabolism" (genetic or ontogenetic), the absence of disease, parasites, and other physically debilitating conditions (diarrhea, for example, is a major problem in third world countries), and such directly health-related variables as muscle tone, cardiovascular conditioning, and personal hygiene. I fully appreciate that many factors can influence physical health (nutrition, pollution, public order, working conditions, sleep, etc.). However, I have assigned these health-related factors to our other categories of primary needs; I want to adhere closely to the principle that each primary need category should contain irreducible, nonredundant elements of the overall adaptation problem.

Mental Health

Including mental health as a primary need might seem questionable to some readers, except that biological adaptation also implies the capacity of any organism to engage in productive, life-sustaining activity. So mental health is not used here in relation to personal fulfillment, happiness, or a carefree existence. Rather it refers to a "state of mind" that allows an individual to carry on normal functioning and self-care without significant impairment (harm). There is a very large research literature on various cognitive, mental, and even emotional dysfunctions in animals and humans alike. Furthermore, in the case of a complex social animal like *Homo sapiens,* the concept of mental dysfunction extends to more subtle aspects of individual psychology like self-esteem, emotional stability, and social integration and status (or the opposite, social isolation). The serious problem of posttraumatic stress disorder in soldiers returning from the Iraq War is just one well-publicized current example. In the past, our perspective on this important aspect of human behavior was racked by polarized attitudes and bitterly competing schools of thought (and methodologies). On the one hand, Sigmund Freud, the founder of psychoanalysis, argued that neuroticism is inherent in society owing to the misfit between human nature and the unnatural demands and constraints of civilization.[33] At the other extreme, skeptics like Thomas Szasz have argued that mental illness is a myth—a syndrome fabricated by therapists and supported by the tendency of a society to label as "sick" any behaviors that are deviant or eccentric.[34] However, during the past twenty years or so a new consensus seems to have emerged to the effect that mental illness is very real; it is cross-cultural in nature; it assumes many forms; and it is affected by a great many causal factors, both biological and environmental. Moreover, the consequences for mental functioning can range from mild anxiety or minor cognitive dysfunctions to total incapacitation and even death.[35]

Communications

The ability to transmit and receive information from the environment, including "feedback" in the strict sense of the term, is a fundamental property of all complex organisms and is an absolutely essential tool in ontogeny (development), adaptation, and reproduction alike. Moreover, the scope of this need is even greater in a social species, where an individual's behaviors must very often be coordinated with others' to accomplish many of the tasks associated with the survival enterprise. Unfortunately, "information" is used in many ways; potentially it could refer to an indiscriminately vast domain—an unmanageable, and unmeasurable, diversity of processes and technologies. Here I confine the definition to what I call "control information"—the capacity (know-how) to control the acquisition, disposition, and use of matter/energy in purposive activities. (For a detailed discussion of this concept, see my *Holistic Darwinism.*) Control information refers to what an individual organism needs to know or communicate to others in order to manage personal adaptation. It includes the information needed to control the activities and events related to meeting instrumental adaptive needs in a given situation. These informational needs are also highly context-specific. They may or may not include literacy or higher education, cooking or computer skills, work skills or social skills. How do we know when "harm" has occurred with respect to this basic need? The answer is, whenever the lack of information or communications skills significantly affects a person's ability to meet other primary needs (to carry on normal functioning). An obvious example is a lack of knowledge about basic hygiene, with the result that the person may be much more susceptible to debilitating diseases or parasites. Likewise, a lack of knowledge about some physical threat (say poisonous mushrooms) can have fatal consequences. (Another, perhaps familiar example is the experience of being in a foreign country without knowing the language; the pronounced feeling of helplessness may be relieved only when an interlocutor is found who can communicate in our own language.)

Social Relationships

This is another primary need that could be interpreted so broadly as to encompass virtually all of our interactions with others. To complicate matters, there is increasing evidence that our social needs are at once critical to the ability to engage in normal functioning and an intrinsic psychological motivation, an evolved aspect of "human nature." Deficiencies in this domain may also have serious consequences for our mental health, our physical health, or both. In addition, social relationships are a primary means of gaining access to information in any social context, as well as providing the cooperative social structure within which we pursue many of our survival-related activities—from production to reproduction. (Recall our earlier discussion of "social capital.") Indeed, our social needs begin at birth. They play a key role in child development, and they remain critical throughout maturation. The analytical challenge associated with measuring this primary need, then, is how to narrow the scope to focus on its role in adaptation and its potential for doing "harm" to our continued capacity for survival and reproductive success. The precise forms these relationships may take differ significantly from one society to another, but there are certain common elements. These include stable, supportive, caring relationships with parents or other closely associated adult caretakers, acceptance by and supportive interactions with peers, and a "positive" social environment, meaning that the operative social values and goals are not alienating, destructive, or exploitative—that is, harmful.

Reproduction

We move now to the first of two primary needs domains that may seem problematic and debatable. They are not usually found on lists of basic needs. (I have broken the process down into two distinct needs, reproduction and child nurturance, because they entail distinct challenges and very different instrumental needs.) Under the heading of reproduction, the focus is on conception and the status of the mother and fetus, along with attendant information, nutrition, health services, and the like. Of course, birthing also involves a distinct set of health risks, services, and skills. However, this need also begs the question: In a world where excess population growth may in fact seem to be a threat to our survival—a part of the problem—how can we justify including reproduction and child nurturance among our primary needs? One reason is that they are absolutely essential from a long-term perspective. Nature has made reproduction an integral part of adaptation for virtually all living species, like it or not, and a significant portion of our collective activity as a species is devoted to reproduction in all its facets. Indeed, the world population problem is a result not of reproduction per se but of too much reproduction—an excess over what is needed to sustain ourselves as a species. However, two major questions are raised by including reproduction and child nurturance as primary needs. One is, How do we interpret a failure to reproduce? Does this mean the individual—male or female—is maladapted? In a strictly biological sense, the answer is yes (at the individual level). And many people are acutely disappointed if they are unable to produce children. However, this is preeminently a primary need that can also be viewed from an aggregate, population-level perspective. Even if many individuals in a human society do not reproduce, the population as a whole may be well adapted if it is able to reproduce itself over successive generations. Indeed, many nonreproducing individuals may nevertheless contribute in various ways to the successful reproduction of a population. The second question arises out of the fact that a conflict may occur between the primary needs of the parents and those of their offspring. The sociobiological term parental investment can involve a zero-sum relationship in which reproduction and child nurturance require a sacrifice of parental needs and of parental adaptation. How are these trade-offs to be commensurated? The answer in strict biological adaptation terms is that it is likely to be more adaptive if parental sacrifices can be minimized. Indeed, the prolonged period of childhood dependency and the complex nurturing needs of human children put a premium on long-term parental health, competence, and support. So the adaptation of parents and that of their offspring are not easily decoupled. However, if a choice must be made, the biological/adaptive paradigm favors the children—especially if it ensures the "magic number" of 2.1 offspring on average (the replacement rate).

Nurturance of Offspring

From a biological perspective, the last of our primary needs is not merely an afterthought or an easily compartmentalized subtask. It is the culmination of the process of adaptation. In a very real sense, it has to do with the investment we make in our biological future. I am referring, of course, not to our conscious, culturally shaped goals and values but to the ultimate goal of the survival enterprise, and the associated imperatives. Accordingly, child nurturance is a primary need that is "more equal" than any of the others (to paraphrase George Orwell again). It entails every one of the other primary needs—both for the children and for their caretakers—during a period of dependency that may be confined to the first few years or may persist well into adulthood, depending on the particular culture and the economic "niche" that is occupied within that culture. Not only do young children have the same array of physical/material needs that apply to adults—thermoregulation, food, water, mobility, and so on—but they have very specific developmental needs—cognitive, social, and emotional needs (and skills) that are increasingly well understood. Some of these nurturing needs may be provided for through shared goods (heat and shelter are common examples); others may be provided for at the margin (investments of time in child care); but other nurturance needs unavoidably require the provision of economic surpluses, or trade-offs. Indeed, outright conflicts may arise between the needs of the next generation and the needs (and wants) of the present generation, and any society (and its political system) that makes de facto choices in favor of the older generation is maladaptive in the strict sense. Unfortunately, our record (historically) as a species has been uneven in meeting the need for child nurturance. Witness the data worldwide on infanticide, avoidable infant mortality, child neglect and abandonment, and even child exploitation and slavery. Any society that does not make nurturing the next generation a primary objective is jeopardizing its future.

*

It should go without saying that our basic needs, and the means we use to satisfy them, encompass most of what goes on in any complex economy. Lurking just inside the "revealed preferences" that economists prefer to deal with are our basic biological needs, the very core of human nature. To illustrate, the survey of consumer spending conducted by the Bureau of Labor Statistics in 2000 found that 90 percent of our expenditures (on average) fall into seven categories, all related to our basic needs, from housing (32 percent) to transportation (20 percent), food (14 percent), health care (5 percent), and clothing (5 percent).[36] Our fourteen primary needs, and the instrumental means required to satisfy them in a given context, thus provide a solid foundation for evaluating biological adaptation in human societies.

Equally important, this framework provides the theoretically grounded definition of basic needs that Sen and others have called for. It provides a consistent basis for defining and measuring basic needs and their satisfaction in any given society. And it predicts that the denial, or serious deprivation, of any one of these needs will cause significant "harm" to an individual's chances of survival and successful reproduction.

So how does one go about measuring basic needs satisfaction? If the problem of defining our basic needs is anything but simple and straightforward, measuring them is an even greater challenge. In fact, there are a number of ways of measuring needs satisfaction, depending on the analytical focus, the analyst's objectives, and various practical data-gathering considerations. In addition, there are many complex measurement and validation issues, especially where instrumental needs are involved. What constitutes adequate shelter, for instance? Or an appropriate level of education and training? Or sufficient income? And who should make the call on these issues—the individual or some outside expert using bureaucratic or technical criteria?

None of these problems is terra incognita, however. As I mentioned earlier, there already exists a very large body of social intelligence—the fruit of many years of research and development by many researchers and organizations. A great many ongoing data-collection efforts are already in place, and a broad array of useful social indicators are currently in use. Particularly notable are the "poverty indicators" published annually by the World Bank, which currently cover about two hundred economies worldwide. Other important sources of data include the Food and Agriculture Organization of the United Nations (FAO), the World Health Organization (WHO), the United Nations Development Program (UNDP), and a number of national-level programs, especially in the Scandinavian countries. In the United States, the Departments of Agriculture, Commerce, Education, Health and Human Services, Housing and Urban Development, and Justice, as well as a wide range of nongovernmental agencies, collect data that are relevant to basic needs satisfaction as this term is defined here.

The UNDP program is especially noteworthy. The UNDP annually publishes a series of global human development measures, including a Human Development Index (HDI) and a mirror image of it called the Human Poverty Index (HPI), which Amartya Sen had a hand in developing.[37] These indexes are severely constrained by the difficulties of obtaining relevant data in many countries. Nevertheless, they represent quite useful "outcome" measures. The HDI measures longevity (life expectancy at birth), knowledge (adult literacy), and the standard of living (GDP per capita). (In the most recent update, Norway ranked first while the United States is now an embarrassing tenth on the list.) The HPI, by contrast, represents an overall measure of progress (or lack of it) in achieving human development. It measures rates of premature death, levels of illiteracy, lack of improved water sources, and the percentage of children under five who are underweight.

Consequently, many survival-relevant indicators already exist, encompassing many of our basic needs. These statistics include, among others, calorie consumption levels, the incidence of malnutrition, access to safe drinking water, availability of sanitation facilities, poverty levels (using various standards), unemployment, work-related injuries and illnesses, access to health services, immunizations, the incidence of violent crimes, schooling, and such sensitive health-related statistics as infant and maternal mortality and life expectancy. Unfortunately, the implementation of these measures in the less developed countries has been spotty at best.

Also encouraging is the recent revival of interest in developing social indicators under the banner of national "well-being." An International Commission on the Measurement of Economic Performance and Social Progress established by the president of France, Nicolas Sarkozy, in 2008 recommended that governments and economists should move beyond reliance on GDP as the primary measure of progress by monitoring household income and consumption, along with various measures of health, education, and environmental conditions. As the commission's final report observed, "Quality of life depends on people's objective conditions and capabilities." The commission certainly had the right diagnosis, but unfortunately the members could not agree on a prescription. They recommended only "further research."[38]

What can the Survival Indicators paradigm add to these efforts? First, it provides a theoretical framework for ordering and rationalizing various existing indicators in terms of the underlying biological survival problem. Second, it expands the horizon of the existing body of social and health indicators to include some additional areas of concern that are often slighted (e.g., thermoregulation, mobility, and mental health) or that are typically defined and measured in rather narrow terms (e.g., freshwater supplies and sanitation).

But most important, the Survival Indicators paradigm has a broader, and deeper, objective than simply to provide better statistics for monitoring the social correlates of economic development, or social well-being, or optimum personal development, however important these objectives may be. The Survival Indicators paradigm addresses a question that many social theorists are not even asking, though they should be: How are we doing in terms of the basic survival problem? This question, in turn, implies a multilevel, multifaceted approach to measurement. Adaptation can be addressed at the individual level or at the population level; it can be directed to the primary needs level or to the provision of instrumental needs; and it can focus positively on documenting needs satisfaction or negatively on the evidence of "harm"—that is, failures in meeting basic needs. Perhaps most important, the Survival Indicators framework is focused directly on outcomes, not on various surrogate measures. (All of this is discussed in greater detail in *Holistic Darwinism.*)

Implicit in the Survival Indicators framework is a major shift in the way economic, social, and political phenomena are viewed and analyzed. As I suggested above, the ongoing survival and reproduction problem, and the basic needs associated with it, apply to all of us in every society. Moreover, much of our economic activity is devoted to meeting these needs, even when we label some product a "luxury" item. Fur coats, after all, do serve a primary human need—they keep the wearers warm. (Of course, many substitutions for fur coats are available today, but for some of our remote ancestors living in high latitudes or at high altitudes, fur coats were nonsubstitutable instrumental needs.) In a similar vein, king-size beds enable us to satisfy our primary need for sleep, even though less imposing accommodations may serve just as well.

From a biological perspective, our primary needs provide the inner logic of economic life. They represent the agenda that implicitly shapes our economies, and it is possible to view all of economic, social, and political life in terms of their relation to the survival imperatives. As I have suggested, much of our economic activity is in fact instrumental to our survival; it is either directly or indirectly related to the satisfaction of our biological needs. To be sure, some economic activity is tangential or not related at all. As I noted earlier, some activities may even be destructive to our adaptive needs—as Robert Edgerton documented in depth in *Sick Societies*. Smoking and hard drugs are obvious examples, but so is almost anything that is carried to extremes—for the simple reason that our survival and reproductive needs are manifold; if we satisfy any one of these needs to excess, we may well jeopardize other needs. (For a book-length treatment of this issue, see Frances Ashcroft's *Life at the Extremes*.)[39]

Many insights about economic, social, and political life can be gained by viewing them from an adaptation/fitness perspective. For instance, it might shed further light on such traditional economic concepts as "discretionary income," "demand elasticity," and the logic of "substitutability." But more important, our biological needs create economic imperatives that allow us to formulate many "if-then" predictions about our economic choices and behaviors. Many of these predictions already make intuitive sense to us. For instance, we can predict in general (but not in every detail) what would happen if the water supply for a major metropolitan area (say the reservoirs that serve San Francisco) was suddenly, irreversibly contaminated by a terrorist. Likewise, we can make predictions at the individual level about how a person's priorities will change as a consequence of the prolonged deprivation of any one of their primary needs (excepting possibly reproduction and child nurturance).

As a thought experiment, just imagine how difficult it would be to continue working or studying in the face of an extended denial of such primary needs as sleep, food, water, waste elimination, or heat (on a frigid day). These deprivations happen, often enough, and they produce predictable consequences. Moreover, most of us spend most of our available time and energy on activities that are directly or indirectly related to satisfying our basic needs. (The American Time Use Survey conducted by the Bureau of Labor Statistics indicates that the pattern in the developed countries is not as different from many less affluent societies as is commonly believed, though we spend more of our leisure time watching television and using various electronic devices.)[40] To deny the relevance of our primary biological needs in our daily existence and in shaping our priorities for the future is to deny reality.

One of the major challenges for the science of economics, then, is to use the biological adaptation perspective as an analytical framework. How well is any given individual, or our society as a whole, doing with respect to the fourteen domains of basic needs? This, in turn, implies a revisioning of our assumptions about the underlying purpose of human societies and the biological consequences of economic life.

Such a revisioning presents an important analytical opportunity. But more important, at this critical juncture in our history it is also an increasingly urgent moral imperative. As Socrates (Plato) truly noted, we cannot adequately define social justice without reference to our basic needs. Yet, in different ways, both capitalism and socialism fall short in their understanding of human nature and in their stance toward social justice. It's time to confront these two outdated nineteenth-century ideologies.

6 * *Why Capitalism and Socialism Are Unfair*

> Justice is entirely empty and meaningless. . . . [Any claim to the contrary] is either thoughtless or fraudulent.
>
> FRIEDRICH HAYEK

> Property is theft. . . . Property is despotism. . . . [Socialism is] an aspiration towards the amelioration of society.
>
> PIERRE-JOSEPH PROUDHON

Darrell Huff's humorous little 1954 book *How to Lie with Statistics* has become an American classic, and rightly so. A reviewer for the *Atlantic* magazine called it "a subversive book, guaranteed to undermine your faith in the almighty statistic." With the artful use of some memorable examples, Huff amply confirmed the remark attributed to British statesman Benjamin Disraeli: "There are three kinds of lies: lies, damned lies, and statistics."

To be sure, statistical methods can be a powerful tool for illuminating otherwise obscure relationships, patterns, and trends. (Full disclosure: I also use statistics—though with care—most recently in the previous chapter; and many more will be found below.) But statistics can also be used—sometimes with deliberate intent—to distort or mask the truth, shrouding a lie in spurious precision and a false appearance of objectivity. Statistics often come disguised as hard "facts" when they may be only uncertain estimates, or averages.

Exhibit A, sad to say, is the "American dream." Even before our current devastating recession, our national self-image as an affluent land of opportunity and still the richest country on earth—by some measures at least—had become woefully out of date. But the dream lives on because, among other things, some of our official economic statistics have seriously misrepresented the truth. It is what Winston Churchill—with tongue in cheek and in a very different context—called a "terminological inexactitude."

Take unemployment. It's no great secret that this key measure of our economic well-being has been defined so narrowly by the Bureau of Labor Statistics (BLS) that it excludes vast numbers of people who are in fact unemployed or seriously underemployed. The official figure—a rapidly moving target these days that currently stands at about 10 percent—

understates by about half the actual number of people who desperately need work to meet their basic survival needs. The BLS reports that about 15 million people are currently unemployed (as this book goes to press). This is bad enough, but there is also an obscure BLS table, called "Alternative Measures of Labor Underutilization," that includes the many unemployed workers who have become discouraged and have stopped looking for jobs, and the workers who hold only part-time jobs because they cannot find full-time work. The BLS currently puts this more realistic figure at 17.2 percent of the workforce, a vast army of at least 25 million people (and their children) whose basic needs are not being met, and even this is surely a gross undercount. (For example, this statistic excludes any estimate of the many potential new workers who have never been employed, since our economy needs to create about 125,000 new jobs each month to keep up with population growth.)[1]

A similar "understatement" exists with respect to another key economic indicator—the "poverty line." This closely watched figure—derived from an elaborate formula for estimating the cost of living—has been used since the 1960s to determine how many people in this country are currently living in poverty. Before the housing and financial bubble burst, the official poverty line statistic for 2008 was 13.2 percent of the population, or 39.8 million people.[2] Today the number is likely to be closer to 50 million. But even this is a low-ball estimate.

The fact is that there are many problems with the poverty line indicator, and it has been subject to much critical debate and analysis over the years. One concern is that any family earning even one dollar more than the current income ceiling is by definition not in poverty. This is arbitrary and misleading.

But the most serious problem is that the formula used for calculating the poverty line is unrealistic. At least twenty independent studies over the past decade or so have come to essentially the same conclusion.[3] The poverty line understates by about half the amount of income a person, or a family, needs for even a minimally decent standard of living (especially if you use the fourteen basic needs domains enumerated in chapter 5). It is based on an annual income (in 2009) of less than $22,050 for a family of four. But a more realistic figure is probably closer to $40,000, with some variation to allow for differences in the cost of living in different regions. In other words, the number of people living in poverty in this country is really closer to 25 percent of the population, or about 75 million people. (This is consistent with the results of a Gallup Poll a few years back in which 27 percent of the respondents characterized themselves as poor.)[4]

Many other indicators point to the same conclusion: America has become a land of widespread poverty and hardship. We have been falling behind other industrialized nations. Here are just a few bullet points:

- Several studies in recent years have estimated that about 30 percent of those who are fully employed are "working poor," many of whom have to hold down two or more low-wage jobs to make ends meet. In a TV news interview for PBS, one struggling mother of two who was juggling three part-time jobs commented, "For me, the American dream is just being able to survive."[5]
- Even those who still have better-paying jobs must work much harder than their parents and grandparents did. Back in those days, a middle-class lifestyle could very often be sustained with only one-wage earner in the family. Today it typically requires two. Indeed, the average household puts in 26 percent more work hours than was the case thirty years ago, and full-time American workers logged an average of 212 more hours during 2007 than workers in the Scandinavian countries (the equivalent of 5.3 more workweeks). The result of all this extra work, furthermore, amounted to doing more for less. Harvard law professor Elizabeth Warren, a student of middle-class economic issues, has concluded that "never before have families worked so hard just to break even. . . . We estimate that the total fixed costs for a typical family with young children, including taxes, mortgage, child care, transportation, and health insurance, has increased by a whopping 165 percent over the past generation. . . . Even a solidly middle-class family has less money for food, clothes, school expenses, entertainment, and–most significantly–savings and long-term wealth building."[6]
- In 2007, 36 million people in this country were estimated to have experienced "food deprivation" (hunger) at various times during the year, including 16.9 percent of all children. Now the number is closer to 50 million, including 17 million children.[7] Food stamp purchases are also at an all-time high at about 36 million participants as of 2009, at an estimated cost for the year of about $50 billion.[8]
- Before the recent, long-overdue increase in the minimum wage in this country, it had remained at $5.15 an hour for ten years, from 1997 to 2006, which amounted to a cumulative decline in its purchasing power over the years of almost 20 percent. But even the current minimum of $7.25 an hour is patently inadequate as an income "floor." Consider this: If you were to work a forty-hour week at the minimum wage and had no deductions at all, you would take home (if you had one) a total of $290 per week, or $15,080 per year. In other words, your earnings would still be far below the poverty line for a family of four. This is why some states mandate minimum wages higher than the federal level, and why

some localities have implemented "living wage" ordinances that typically start at $10 an hour or more.

- The number of people without health insurance in this country—an international embarrassment, since we are the only developed country without universal coverage—was closing in on 50 million uninsured people before the health reform legislation was finally enacted, while many millions more had inadequate coverage. The consequences of this national disaster have been showing up in our basic health statistics. We now rank forty-fifth among the nations of the world in infant mortality, below such countries as Cuba, Slovenia, Greece, Portugal, and the Czech Republic, and our life expectancy at birth is even worse.[9] We are ranked fiftieth behind such unlikely places as San Marino, Monaco, Liechtenstein, and Cyprus, as well as every other developed nation.[10] Significantly, there is also a difference of 4.5 years in average life expectancy between the bottom and top 10 percent of the population in terms of income, up from 2.8 years in 1980.[11]

Franklin Roosevelt, in his second inaugural address in 1936—in the depths of the Great Depression—declared, "I see one-third of a nation ill-housed, ill-clad, ill-nourished." The sad reality is that his words still ring true today, and it is high time for us to face up to it.

But this is only half the story. While the poor have been getting poorer, and more numerous, over the past thirty years—since the beginning of Ronald Reagan's "morning in America" presidency—the rich have been getting richer, and richer, and richer. During the George W. Bush years, especially, there was a gusher of new wealth at the very top of the economic pyramid, much of it at the expense of the poor and the working class. As journalist Barbara Ehrenreich pointed out in her disturbing "undercover report" (she posed as a low-wage worker) called *Nickel and Dimed: On (Not) Getting By in America,* the working poor are in fact major philanthropists in our society. "They neglect their own children so that the children of others can be cared for; they live in substandard housing so that other homes can be shiny and perfect; they endure privation so that inflation will be low and stock prices high. To be a member of the working poor is to be an anonymous donor, a nameless benefactor, to everyone else."[12] Consider just a few indicators of this philanthropic sacrifice:

- Since the 1980s, some 94 percent of the total increase in personal income has gone to the top 1 percent of the population (about 2.7 million people). According to the *Economist* newspaper, in 2000 the top 1 percent of the population received 20 percent of the total income for the year, while the bottom 80 percent had 41 percent.[13]

- In 2001, the top 1 percent of households owned about 33.4 percent of the total personal wealth in this country (including housing but excluding cars, furnishings, and personal items), while the top 20 percent held 84 percent. The bottom 80 percent combined held only 16 percent of the personal wealth.[14] (It is cold comfort to know that global disparities in wealth are even worse. A recent UN study found that the top 2 percent of the population worldwide owned about 50 percent of the wealth while the bottom 50 percent had a trivial 1 percent.)[15]
- Much has been made about the spread of stock ownership in recent years, thanks mainly to the rise of 401(k) retirement accounts as replacements for the traditional company-sponsored pension plans that have been rapidly disappearing. But the reality is that the top 20 percent of households own some 89 percent of all stocks and bonds, while the bottom 80 percent hold about 11 percent (as of 2007).[16]
- As we all know, CEOs' salaries have ballooned well beyond mere greed over the past two decades. In the 1940s, CEOs' salaries for large companies were about twenty times that of the average worker, and American companies were very well managed on the whole. During the past thirty years, the salary gap has progressively widened to about 260 times that of the average worker, with some CEOs' salaries exceeding 500 times as much.[17] And this doesn't count the plethora of perks that many CEOs have received in recent years—such as liberal use of a company jet, accommodations at company-owned apartments and resort homes, country club dues, "financial planning" services, special executive pensions, low- or no-interest loans (sometimes later "forgiven"), and even having the company pay some of their personal taxes. As one prominent executive compensation consultant explained before the economic collapse and the rise of public anger against corporate greed, "People have a certain pride in pigging out. . . . It's a measure of their worth."[18]

In sum, capitalism has not been working as advertised. Going back to Adam Smith and especially to such early economists as John Stuart Mill and Alfred Marshall, the expectation has been that capitalism would progressively improve the "general welfare." A rising tide would lift all boats. But what if many of the boats have holes in their hulls? What if they are sinking even as the tide is coming in? Or what if the tide is actually going out, as we have observed in this recession?

A more contemporary metaphor, associated especially with the failed promise of supply-side economics in the 1980s, is that new wealth at the top would "trickle down" to the rest of us. But this has not happened. Instead of a trickling down of wealth, we have had a bubbling up. Indeed, in the wake of the great bank bailout, the disconnect between Wall Street and Main Street has become even wider as investment bankers have returned to paying themselves billions of dollars in bonuses while many millions of newly unemployed workers have been sinking into poverty.

In short, capitalism has failed in its basic (utilitarian) promise to enhance the "greatest good of the greatest number." Like the laissez-faire era in the late nineteenth century (the "Gilded Age") and the "Roaring Twenties" (before the Great Depression), so the Reagan-Bush "free market era" has come as close as any we have had to the capitalist ideal—what economist Samuel Bowles (after Karl Polanyi) calls "utopian capitalism." So capitalism's failings cannot be attributed to too much regulation, or confiscatory taxes, or a socialist redistribution of wealth that has stifled initiative and investment. There are no scapegoats! So what went wrong? The answer, in a nutshell: utopian capitalism.

Capitalism in practice is an evolved, highly complex system of social cooperation, competition, and conflict that is ever-changing and still inadequately understood even by professional economists. At the same time, it is a high-profile "ideology"—a set of interconnected concepts and values that map rather poorly onto the way real economies work. Nevertheless, this idealized model has been assiduously promoted by its true believers over the years and has been routinely used to justify self-serving policies and practices, sometimes with disastrous results. Our current Great Recession—the label of choice for this dark era—is a prime example. (The Nobel Prize–winning economist Joseph Stiglitz accuses his profession of being "free-market capitalism's biggest cheerleader.")[19]

Much of what the capitalism boosters would like to take credit for was, of course, invented thousands of years ago. Recall Plato's discussion about the importance of a division of labor and specialization, and the benefits of free exchanges and economic markets. In its day, ancient Athens was the greatest of all trading states, and its central marketplace, the Agora, was legendary. (Recall also Plato's warning about the concentration of wealth and the potential conflict between the "two cities"—the rich and the poor.)

What capitalist ideology has added to this ongoing economic story is a set of ideas about how a complex market economy works, or should work, and the centerpiece of this idealized model is, needless to say, Adam Smith's "invisible hand." As Smith explained it in *The Wealth of Nations,* "Man is . . . led by an invisible hand to promote an end which was no part of his intention. Nor is it always the worse for the society that it was not part of it. By pursuing his own interest he frequently promotes that of the society more effectually than when he really intends to promote it."[20]

Smith also provided a gold-plated rationalization that capitalist economists have used ever since to justify unvarnished greed and an unfettered pursuit of self-interest: "In spite of their natural selfishness and rapacity . . . [men] are led by an invisible hand to . . . advance the interest of the society." Modern economists often become lyrical about "the superiority of self-interest" over altruism in economic life and the virtues of competition and the "profit motive," while overlooking the fact that Smith's rendering of the invisible hand was quite contingent. As he said, the invisible hand is not "always the worse" for society and "frequently promotes" the general welfare. But this is not a sure thing. Many of Smith's acolytes also seem unaware of the cautionary warnings both in his masterwork and in his earlier work *The Theory of Moral Sentiments,* where (as a Stoic and a Christian) he stressed that everything in a free market depends on a moral foundation of trust, honest dealing, and, as he himself put it, "justice." Indeed, Smith was also proponent of the Golden Rule. (See endnote 21.)[21]

The classical economists who followed in Smith's footsteps embellished his core vision in various ways. One of the most important of the early economists, Léon Walras, claimed that the forces of supply and demand if left alone would work to ensure the efficient use of resources, full employment, and a "general equilibrium." In other words, competitive free markets can be depended on to be self-organizing and self-correcting, and the profits that flow to the property owners—the capitalists—will generate the wherewithal for further growth and, ultimately, the general welfare. The modern economist Robert Solow summed up utopian capitalism (generally referred to as the neoclassical model) as a compound of "equilibrium, greed and rationality." (For more, see the endnote.)[22]

In this model, "efficiency" becomes a paramount value. The objective is to use "scarce resources" as productively as possible, presumably to advance the general welfare. However, there is much talk in economics textbooks about the need for a "trade-off" between efficiency and equality, with the implication that the basic needs of labor may need to be sacrificed for the sake of economic growth. When economists define efficiency, they tend to invoke Pareto efficiency, doing whatever is needed to make somebody better off (read capitalists) while—in the words of William Baumol and Alan Blinder in their introductory textbook—"not worsening the lot of everyone else" (read labor).[23]

But this is disingenuous. An efficiency/equality trade-off means the workers are by definition held back or made worse off in the name of efficiency. (This supposed trade-off also happens to be a straw man, because only die-hard socialists are concerned about economic "equality." The real issue is our basic needs!) Also, as economist James K. Galbraith points out in *The Predator State,* "The setting of wages and the control of the distribution of pay and incomes is ultimately a social, not a market decision. . . . Society decides what the distribution of pay should be. . . . There is no trade-off in a properly designed economic policy between efficiency and fairness."[24]

Perhaps most pernicious is the way capitalist economists treat workers simply as an "input" to the production process and not as stakeholders in the economy with rights that complement the claims of property ownership. Economist John B. Taylor, in his introductory textbook, tells us that "labor is what is bought and sold on the labor market." It is a "factor of production" that is subject to labor market forces.[25] Thus Taylor and other textbook writers, like Baumol and Blinder, criticize the minimum-wage law because, they claim, it prevents workers who would be willing to work for less from doing so. They believe the minimum-wage law actually causes unemployment.

Equally troubling is the fact that some economists seek to justify unemployment. As one economist explained to me, unemployment is "the crucial basis of a functioning economy as it is needed for labor mobility." In fact, unemployment is certainly not "needed" at a time in our history when more than 25 million people are out of work or underemployed and when hundreds of millions of Chinese and Indian workers provide ample low-cost labor mobility for American companies (one reason that there have been so many job losses in this recession and that so many of these jobs won't ever come back).

For more than two decades, many economists and policy makers adhered to Milton Friedman's theory that there is in fact a desirable or optimal level of unemployment. The so-called nonaccelerating inflation rate of unemployment (NAIRU for short)—ultimately pegged at about 6–7 percent—was widely viewed as being essential to avoid causing inflation. If the labor market is tight, so the NAIRU theory claimed, workers would bid up their wages and start a wage-price spiral. Therefore some unemployment actually serves the greater good, or so the reasoning goes.

As Galbraith points out, "The idea that low unemployment generates runaway inflation is an absurdity on its face. If it had been true, runaway hyperinflations should have been common in history, whereas in fact they are very rare."[26] Galbraith notes that Alan Greenspan, as Federal Reserve chairman during the later 1990s, allowed unemployment gradually to fall to the point where full employment was achieved, and it did not produce inflation. Just by doing nothing about low unemployment, Greenspan disproved Friedman's theory.

The willingness of economic theorists to endorse unemployment in the name of the greater good is an example of the phenomenon, mentioned in chapter 3, of "emotional disengagement." It leads people to deny the human consequences of their actions. Indeed, Léon Walras recommended that employers pay the lowest possible wages. And David Ricardo's "iron law of wages" predicted that when there is a surplus of laborers and thus "wage competition," wages will naturally fall to a subsistence level if allowed to do so.[27] Adam Smith himself pointed out that "masters are always and every where in a sort of tacit, but constant and uniform combination, not to raise the wages of labour above their [prevailing] rate." Furthermore, Smith noted, "Masters, too, sometimes enter into particular combinations to sink the wages of labour even below this rate."[28] (Recall Plato's "two cities." Marx called it class warfare.)

So, in effect, some capitalist theorists predict and even seem to condone poverty. As economist John Gowdy candidly acknowledged, "Economic theory not only describes how resources are allocated, it provides a justification for wealth, poverty, and exploitation."[29]

Mainstream economists would no doubt respond that efficiency is an essential ingredient for growth and profitability: the benefits of economic activity must outweigh the costs. That's why worker "productivity" measures are so widely monitored in this country, for instance, and why technology improvements are so important. Technology is preeminently a way of leveraging the efficiency of labor. But if a business achieves efficiencies by paying poverty wages, that is another matter. It amounts to a deliberate cause of "harm" as I defined it in chapter 5. Paul Samuelson, the dean of twentieth-century American economists, summed up his personal view of the issue this way: "Never think about [efficiency] without at the same time contemplating its differential effects on different people's well-being. . . . A good democracy is justified in trading off some efficiency against the goal of social harmony or equality."[30]

In reality, our minimum-wage law provides an economic floor for many workers, however inadequate. If employers need more workers, they will grudgingly pay the minimum wage so long as everyone else is forced to do so and will either pass along the costs to their customers, or reduce their profits, or find other efficiencies. Some economists might counter that businesses can (and do) get around the minimum-wage law by outsourcing the jobs to developing countries or by cheating. (In fact, a recent major survey of low-wage workers in several United States cities found that about one-quarter of them were consistently paid less than the minimum wage and that more than three-quarters of those who worked overtime did not receive proper overtime pay.)[31]

If, as some economic theory recommends, American workers have to accept $1 to $2 a day (the going wage in many countries), or accept a subminimum-wage job in order to be competitive with the "next best" (or rather next worst) wage alternative, then our society is in fatal trouble. We will have to find a better solution than this, for any society that sacrifices the basic needs of the population on the altar of profits, or economic progress, or—more sinister—the protection of unlimited property rights for the rich is destined to become politically unstable. Plato's warning should be taken to heart.

This issue also highlights the unrealism of classical capitalist theory. If you assume, as the utopian theorists do, that there is an inherent tendency toward an economic equilibrium with full employment, then there will always be competition for workers in the marketplace, and wage levels will at least be reasonable, if not fair. But what if an economy is dynamic, sometimes unstable, and prone to booms and busts? What if there are in fact chronic labor "surpluses," at least in many low-wage categories, and a shortage of low-wage jobs (in part owing to foreign competition and outsourcing to other countries)? Then workers do not have an option and are in effect being coerced into taking whatever wages are offered, even if they are inadequate or even illegal.

This, in essence, is the plight of the working poor in our society—and in many others as well. And this is precisely why a mandated wage floor, properly enforced, is a necessary corrective. As Bill Gates Jr. (lately the richest man in the world) noted in a TV interview with Bill Moyers a few years back, "Markets only work for people who have money." (For the record, the Bill and Melinda Gates Foundation is the world's largest philanthropic enterprise.)

Another senior economist, Samuel Bowles, points out in his critique and revisioning of economic theory with the unassuming title *Microeconomics* that capitalist doctrine offers "an odd utopia." Its strongest claims are generally false; it is unable to make reliable predictions; it removes from its models many of the factors that shape real-world economies; it relies on simplistic models; it ignores the pervasive and inescapable influence of wealth and "power" in shaping how real economies work; and not least, it's profoundly unfair. It systematically favors capital over labor, with results that are evident in our economic statistics.[32]

In fact, a modern capitalist economy can become a rigged game that is far removed from the idealized market model of individual actors with equal power and resources who rationally pursue their self-interest by engaging in mutually beneficial (win-win) exchanges that produce "efficient markets" and optimal outcomes for all concerned. Among other things, the vast differences in wealth, power, and information among the "players" exert a highly coercive influence on the marketplace and on our political system. This is clearly evident in our often corrupt lobbying system, in the revolving door between our regulatory agencies and corporate America, and not least in the way we finance our election campaigns.

In his systematic indictment of American capitalism in *The Predator State,* Galbraith calls our reigning free market ideology a "myth"—"something that is repeated to schoolchildren [and the voters] but hardly taken seriously by those on the inside." The underlying reality includes "the systematic abuse of public institutions for private profit or, equivalently, the systematic undermining of public protections for the benefit of private clients." (Stiglitz calls it a "corporate welfare state.")[33] Nor are our markets actually "free." They are dominated, and manipulated, by huge corporations, with insufficient checks and balances. Recall Robert Reich's characterization of this as "supercapitalism." Indeed, the conservatives on the Supreme Court recently opened the floodgates for corporate money in politics by reversing generations of law and precedents, including seven previous Supreme Court rulings.[34]

To cite one of the more notorious examples of corporate malfeasance, American cigarette manufacturers spent a generation denying the relentlessly accumulating evidence that cigarette smoking is a major cause of lung cancer and other illnesses, all the while continuing to pursue heavy promotional and advertising campaigns and effectively blocking any legislation designed to curtail this insidious addiction. Fifty years and many laws and lawsuits later, the cigarette manufacturers are now more circumspect and accommodating to various antismoking programs, but they are still profiting from a capitalist enterprise that is clearly "harmful" in the way I defined it in chapter 5.

Another more immediate and disturbing example, because of its implications for our general public health, was glaringly illuminated in the documentary movie *Food, Inc.* Since the 1950s, the food production industry in this country has become heavily industrialized, with huge Concentrated Animal Feeding Operations (CAFOs) and enormous food processing and packaging factories. But what the public is not allowed to see (except with unauthorized hidden cameras) is a lightly regulated production system that brutalizes the animals, the farmers, and the workers alike and, more important, sells food products that put the consumers' health at serious risk. A tip of the iceberg is the increasing frequency and magnitude of *E. coli* bacteria outbreaks in recent years, which have resulted in numerous deaths.[35]

To be sure, modern capitalism comes in many sizes and shapes, from the millions of mom-and-pop businesses with one or a few workers to huge international conglomerates with hundreds of thousands of employees worldwide. But for every Google that provides a cornucopia of perks for its employees, there are many others that are single-mindedly devoted to what I would call an iron triangle of mutually reinforcing values: maximizing growth, maximizing efficiency, and maximizing profitability. Sometimes customer service seems to be the lowest corporate priority. (Another, implicit value is "creative destruction," economist Joseph

Schumpeter's classic way of characterizing the downside cost of economic progress. Though he didn't intend it, Schumpeter's concept has also provided predatory capitalists with a convenient rationalization for ruthlessly crushing their competitors.)

The legendary organic farm innovator Joel Salatin summed it up this way in *Food, Inc.*: "As soon as you grasp for that growth, you're gonna view your customer differently, you're gonna view your product differently, you're gonna view your business differently, you're gonna view everything that is most important differently." The result, all too often, is predatory capitalism.

The notion of equal opportunity and a "level playing field" in economic life is also badly tainted. The book *Unequal Chances,* a major in-depth study edited by Samuel Bowles and two colleagues (Herbert Gintis and Melissa Osborne Groves), documents the systematic bias in what we like to think of as a meritocracy. Among other things, there is a high age-adjusted correlation (0.42) between the income levels of parents and children over time. In other words, the children of affluent parents have a far better than average chance of doing well economically, whereas the offspring of the poor generally do much worse than average. The editors conclude: "Children from the least well off families do not have a fair chance at attaining the level of economic security most other families manage to attain."[36]

Equally important is that, contrary to Adam Smith's dour view of human nature, the capitalist system itself shapes our behavior and, because of its reward structure, favors those who most closely resemble the model of *Homo economicus.* In other words, capitalism has a self-fulfilling aspect. Recall the discussion in chapter 4, where it was shown that human motives and values are far more complex than the simplistic capitalist stereotype would lead us to believe. We are also greatly influenced by "social forces"—the rules, norms, customs, peer pressures, contagion effects, and not least, the "incentives" that are put before us.

In addition, we must recognize the influence of hard-core individual differences—the heterogeneity of personalities and motivations that shape our behavior. One study, cited by Bowles, showed a clear distinction between people with a bias toward equality (22 percent), people who were likely to follow the prevailing norms (39 percent), and those who were self-regarding, even ruthless egoists (29 percent). Another study, by Ernst Fehr and his colleagues, showed that somewhere between 40 and 60 percent of the study participants preferred reciprocity in their relationships, whereas 20 to 30 percent were highly self-centered and insensitive to the needs and rights of others.[37]

Perhaps the most authoritative evidence of categorical differences in our behavioral predispositions comes from the large body of data that has been amassed over more than twenty years by one of the leading career guidance and training organizations, Target Training International. About 30 percent of the many people who have taken TTI's personality assessments over the years show a dominant preference for economic and utilitarian objectives, and another 13 percent are motivated toward political objectives. On the other hand, 14 percent strongly prefer social and humanitarian work, 15 percent favor learning and teaching, and 17 percent have aesthetic and artistic preferences.[38] (Recall also the research in behavior genetics on personality differences cited in chapter 4.)

To repeat, human nature comes in many flavors, but the "free market" capitalist system favors the one-third who are the most acquisitive and egocentric and the least concerned about fairness and justice. So, no surprise, this is the end result that our economy produces. "Greed works," as the character Gordon Gekko claims in *Wall Street* 2, the sequel to a famous 1980s movie. The dynamics of the system are such that they reward ruthless competitiveness and, very often, punish people with a social conscience. If you play fair, you will get wiped out. Intel's legendary CEO and chairman Andrew Grove highlighted this corrosive ethic in his best-selling 1996 book *Only the Paranoid Survive*.[39] Indeed, the not-uncommon corporate mentality of assuming you are "at war" with your competitors can create fierce internal pressures to be loyal to your own "tribe," right or wrong, and to display a steely disregard or even hostility toward everyone else—from competitors to the government and, at times, even your customers.

One of many implications of this business culture (I'll talk about others later on) is that there will always be latent Bernie Madoffs out there in the marketplace. Madoff was unrepentant about what he had done over so many years. He just ran out of suckers when the financial meltdown occurred, so he could no longer sustain his Ponzi scheme. In fact, more failed Ponzi schemes (mini-Madoffs, they have been called) have recently been coming to light. By the same token, it is noteworthy that five out of the twenty top executives of AIG (the insurance giant rescued by the taxpayers) who received undeserved bonuses in early 2009 could not be shamed into returning them. And the *New York Times* reported that nine of the big banks that received a total of nearly $175 billion in taxpayer bailout money that year subsequently awarded $32.6 billion (yes, billion!) in bonuses. Talk about unfairness.

One of those banks, Goldman Sachs, announced record profits later in 2009 and projected year-end bonuses for its 31,000 employees of $16.7 billion. Lord Griffiths, the British vice chairman of Goldman Sachs International, defiantly asserted in a subsequent public speech that the bank had no reason to be ashamed about these bonuses. The public should learn to "tolerate inequality as a way to achieve greater prosperity for all."[40] Unfortunately, the public was not very tolerant. In the furor that ensued, the company decided to change its plans and convert some of the bonuses for its top executives into company stock, a very small concession. And CEO Lloyd Blankfein limited himself to a paltry $9 million!

So we must always expect to have shameless financiers and other opportunists, as well as unscrupulous speculators, sharp lawyers looking for loopholes, and outright swindlers who are trying their best to beat the system or break the rules without getting caught. And any capitalist ideologue who naively assumes that the economic system can be self-policing is just as utopian as the socialists who assume that humans are naturally good, if only the economic system can be reformed.

However, the most fundamental objection to capitalist economic theory is that it is in denial (or at least obtuse) about our basic survival needs. It rejects the very principle of distributive justice (witness the chapter epigraph from economist Friedrich Hayek), and it overlooks the many ways modern market economies create what Bowles calls "poverty traps"—such as the inability of the poor to get low-cost credit or to generate enough savings to escape the disadvantages of rental housing and benefit from homeownership. Then there is the favored tax treatment for capital gains over wages and for property owners versus renters.

As economists Paul Samuelson and Richard Nordhaus concede in their classic introductory economics textbook, markets don't produce "a fair distribution of income and prosperity."[41] Baumol and Blinder, in their textbook, half-heartedly concur: "Many observers also believe that markets do not necessarily lead to an equitable distribution of income." Elsewhere in their volume, though, Baumol and Blinder seek to justify this state of affairs. "Nothing in the market mechanism guarantees equality. On the contrary, the market tends to breed inequality, for the basic source of its great efficiency is its system of rewards and penalties. The market is generous to those who are successful in operating efficient enterprises . . . and it is ruthless in penalizing those who are unable or unwilling to satisfy consumer demands efficiently."[42]

To some, this is redolent of social Darwinism (but not of Darwin's own morally grounded vision of human societies, remember). It converts what was for millions of years a mutually beneficial collective survival enterprise (as described in chapter 3) into a gladiatorial arena, an economic version of the Olympic Games in which the winner gets the commercial endorsement fees and all the rest of us become high school swimming coaches—if we're lucky. In sum, if you have an economic system that encourages and rewards "tough-minded," self-serving behavior with winner-take-all incentives, one that allows you to be exploitative toward workers, indulgent toward cheaters, and even manipulative toward customers, this is exactly the system you will get.

The reality of how a modern capitalist system works has led to what I would call "adjectival capitalism"—or Rube Goldberg capitalism. Instead of a smoothly functioning Newtonian clockwork, we have a combination of "crony capitalism" (e.g., when ratings agencies and government regulators collude with hedge fund managers), "klepto-capitalism" (like Russia after the fall of the Soviet empire), "mafia capitalism" (like the drug lords in various countries), "ersatz capitalism" (a term coined by Stiglitz that aptly describes his experience with the global economists at the World Bank), "casino capitalism" (the term inspired by John Maynard Keynes and used by social scientist Susan Strange to characterize modern financial markets), "permissive capitalism" (economist Paul Krugman's term for the big bank bailouts), and "naked capitalism" (when markets are manipulated, workers are exploited, and small mom-and-pop businesses are crushed). Walmart is the eight-hundred-pound poster child for many people.

Especially important and often underrated by the apologists for capitalism is what I would call "subsidized capitalism." For all President Reagan's talk in the 1980s about government's being the problem and not the solution, the reality is that private businesses seek government assistance wherever and whenever they can—from protective tariffs to export subsidies, special tax breaks, government purchase orders, infrastructure improvements, and even direct research and development assistance. In fact, contrary to the neoconservative mythology, the taxpayers have been a major facilitator for economic development in this country ever since colonial days.

It started even before the American Revolution with the construction of harbors, canals, and "post roads" for mail service. Later on the government subsidized the development of agriculture with land giveaways, agricultural research, farm assistance of various kinds, and (since the 1930s) direct farm subsidies that nowadays distort the food economy and enrich large corporate farmers. Railroad development in this country benefited from land grants that allowed railroads to raise capital by selling land and, not coincidentally, letting the "robber barons" get rich. More recently, the government has built dams and irrigation systems, power grids, highways, and airports, and it has been a major incubator for the development of computers, the Internet, and most recently alternative energy technologies. And this is only a partial list.

Finally, there is what could be called "neo-imperialist capitalism." Stiglitz, in his earlier critique of the globalization process during the 1990s and beyond, *Globalization and Its Discontents,* described in detail how the World Bank, the International Monetary Fund, and the World Trade

Organization (WTO), in return for providing assistance to developing countries, imposed on them what he characterized as "a curious blend of ideology and bad economics."[43] Perhaps most blatant, in Stiglitz's words, was "the hypocrisy of pretending to help developing countries by forcing them to open up their markets to the goods of advanced industrial countries while keeping their own markets protected, policies that make the rich richer and the poor more impoverished—and increasingly angry." Stiglitz estimated that these misguided policies probably pushed an additional 100 million people worldwide into poverty.

No wonder there were protests and even riots at WTO meetings during those years, most notoriously in Seattle in 1999. Stiglitz concluded, "The net effect of the policies set by the [so-called] Washington Consensus has all too often been to benefit the few at the expense of the many, the well-off at the expense of the poor. In many cases commercial interests have superseded concern for the environment, democracy, human rights, and social justice."[44] (Stiglitz also noted the irony that the motto of the World Bank, founded with great idealism at the end of World War II, is "Our dream is a world without poverty.")

Mainstream economists have a ready-made explanation for the various imperfections in the utopian model. They are termed "market distortions" and "market failures." These can include anything from a starving army of unemployed workers to a despoiled environment, trade-killing protective tariffs and import quotas, farm subsidies for rich farmers, and the seemingly irresistible desire of business owners to control or monopolize their markets and control government regulators. (Consider, for example, the oil cartel.) But many so-called market failures really are no such thing. Rather, they represent basic structural deficiencies in the idealized model, and they cannot be fixed simply by tinkering with the machinery, like adding a few more regulations. They are designed to fail. (To be sure, sometimes market solutions can be developed for dealing with "externalities," such as the cap-and-trade system for curtailing greenhouse gas emissions.)

What could be called "bubble-up capitalism"—the recent contagion of greed and reckless behavior that loudly popped when the subprime mortgage mirage suddenly vanished—is not simply a financial crisis or an economic crisis. It's a human crisis, with the loss worldwide of many millions of jobs, an ever-widening swath of poverty, and a growing deprivation of basic needs. Never before in the history of capitalism have so few caused so much harm to so many (to borrow a bit of Churchillian rhetoric). A banner on display at a 2009 G-20 summit meeting in London said it all: "Capitalism Is Not Working."

There is, of course, a raging debate going on among economists these days about what went wrong, why it wasn't predicted (some people did, but they were ignored), and how to fix it. The Nobel Prize–winning economist Paul Krugman, in a 2009 public lecture, argued that most of the theoretical work in macroeconomics over the past thirty years has been "spectacularly useless at best, and positively harmful at worst."[45]

Similarly, in an open letter to England's Queen Elizabeth in response to her public query about why the economic meltdown had not been foreseen, ten of that nation's prominent economists (later supported by two thousand other social scientists worldwide) pointed to fundamental deficiencies in economics as a discipline. "In recent years," these economists wrote, "economics has turned virtually into a branch of applied mathematics, and has become detached from real-world institutions and events." The critics also charged their colleagues with "promoting unrealistic assumptions that have helped to sustain an uncritical view of how markets operate. . . . simplistic and reckless market solutions have been widely and vigorously promoted by many economists."[46] (I would describe it as Panglossian, after Dr. Pangloss in Voltaire's novel *Candide,* who naively believed that "all is for the best in the best of all possible worlds.")

The root of the problem, as various critics have pointed out, is that economists have relied on computer models with assumptions that conform poorly to the way the real world works—such as perfectly efficient markets, risks that can be precisely measured, prices that perfectly reflect all the relevant information, and financing that is always available. As the *Economist* noted, "Economists can become seduced by their models, fooling themselves that what the model leaves out does not matter. . . . They are more wedded to their techniques than to their theories. They will believe something when they can model it." As one mainstream economist expressed it, "If it isn't modeled, it isn't economics." (This brings to mind a saying attributed to Albert Einstein: "Not everything that can be counted counts, and not everything that counts can be counted.")[47]

Just as "war is too important to be left to the generals," to quote the French prime minister Georges Clemenceau's famous remark during World War I, an economy is too important to be left to the economists. Emeritus economist Herbert Gintis is outspoken in his criticism of the discipline. Economists, he says, "consistently choose textbooks that teach material that they know is false and/or completely out of date. . . . They teach stuff that you know is not true. They know the general equilibrium model is not true . . . but they teach it. The production theory that is taught is also a joke. . . . So why do we teach undergraduates this?"[48] Samuel Bowles points out that every respectable economist these days rejects the *Homo economicus* model of human nature, yet they are unwilling to abandon the economic theory based on this assumption. (Economist Richard Thaler has suggested that his profession needs to replace *Homo economicus* with *Homo sapiens.*)

Then there is that stunning mea culpa by Alan Greenspan, who was chairman of the Federal Reserve from 1987 to 2006. Greenspan is a follower of Ayn Rand (you may recall) and an architect of the largely successful effort to immobilize the regulatory powers of the Federal Reserve in the interest of encouraging market freedom and creative "financial engineering." Paul Barrett, an assistant managing editor of *Business Week,* charges that Greenspan "helped shield the mad scientists of Wall Street from government restraints." (Stiglitz calls the Wall Street investment houses "Frankenstein factories.") Much too late, in the fall of 2008, Greenspan admitted to Congress that he had been "shocked" by what had happened. "I made a mistake in presuming that the self-interest of organizations [such as banks and investment houses] were [*sic*] such that they were best capable of protecting their own shareholders." Some mistake![49]

Greenspan and a great many others, from the writers of economics textbooks to legions of bankers, recent presidents from both parties, and even the august International Monetary Fund, were seduced into believing the age-old delusion that the lessons of history no longer applied in our more "enlightened" era. As economists Carmen Reinhart and Kenneth Rogoff conclude in their definitive study of financial crises, *This Time Is Different: Eight Centuries of Financial Folly,* "The lesson of history, then, is that even as institutions and policy makers improve, there will always be a temptation to stretch the limits. Just as an individual can go bankrupt no matter how rich she starts out, a financial system can collapse under the pressure of greed, politics, and profits no matter how well regulated it seems to be."[50]

The respected financier and former ambassador Felix Rohatyn, in a news media interview some years ago, asked rhetorically, "Does the system work to spread the wealth in some way that's reasonably fair?" His conclusion: "Clearly at this point, the answer is no, and that's not tolerable."

As we shall see in the next two chapters, our economic system does not have to work this way. During the World War II era and immediately afterward, the American economy was far more equitable, though it was no less capitalistic. The top 1 percent of families then had about 27 percent of the wealth (versus 38.1 percent by the late 1990s and early 2000s). As for income (including capital gains), the top 10 percent of families in those days received 32 percent of the total (versus 49.3 percent in 2006).[51] Back then, to repeat, CEOs' salaries averaged about 20 times that of the average worker (versus over 260 times today). Immediately after the war, the GI Bill of Rights enabled millions of returning veterans to gain entry into an expanding middle class and buy low-cost homes, and the top marginal tax rate stood at 91 percent! The postwar years were widely touted as "the era of the common man." Yet none of this detracted from what was also a period of robust and dynamic economic growth and spreading affluence.

In other words, we can choose a different, more equitable social contract. Utopian capitalism not only is seriously flawed but also leaves out of the picture much of what goes on in our complex society—from not-for-profit educational, scientific, and business enterprises to volunteer organizations and religious institutions. An economy should not be primarily concerned with growth, or efficiency, or maximizing profits. Its purpose is to provide for the basic needs of the collective survival enterprise in a sustainable way. Entrepreneurship, private property, and capital have an important role to play in this enterprise, but they are not the be-all and end-all. Nor do we have to choose between a "free market" and a government-run "socialist" economy, for the socialist ideal is every bit as deficient as utopian capitalism, though in different ways. So let's take a look at the socialist model.

Socialism certainly has its heart in the right place. Or perhaps I should say "hearts," because socialism is a term that embraces a rainbow of different ideological and political agendas; it's far more heterogeneous than mainstream capitalist doctrine. Although the roots of modern socialism can be traced back to Rousseau's moral outrage against the economic conditions in prerevolutionary France and his vision of a voluntary community of equals, (as described in chapter 4), the modern socialist impulse—and the term socialism itself—was inspired by a similar revulsion against the brutal and exploitative working conditions that accompanied the Industrial Revolution and the rise of the factory system in the late eighteenth and early nineteenth centuries.

"Hard Times," an edited collection of historical documents by various observers of this shameful episode, vividly conveys the slavelike conditions that English factory workers and American sweatshop employees endured during that era—and that still exist today in some dark corners of this country (not to mention in various developing countries). Men, women and even very young children typically worked a fifteen hour day, with only a few short breaks, six to seven days a week in hot, crowded, and dangerous conditions. Accidents and premature deaths were all too frequent. Sir James Kay-Shuttleworth, a physician who worked for many years in Manchester (one of England's leading factory cities), provided us with a firsthand description of those "dark satanic mills," in poet William Blake's memorable phrase:

> The dull routine of a ceaseless drudgery, in which the same mechanical process is incessantly repeated, resembles the torment of Sisyphus. . . . To

> condemn man to such severity of toil is, in some measure, to cultivate in him the habits of an animal. . . . His house is ill furnished, uncleanly, often ill ventilated, perhaps damp; his food . . . is meager and innutritious; he is debilitated and hypochondriacal, and falls the victim of dissipation. . . . Rising before day-break, between four and five o'clock the year round . . . he swallows a hasty meal or hurries to the mill without taking any food whatever. At eight o'clock half an hour [is] allowed for breakfast. In many cases, the engine continues to work during mealtime, obliging the labourer to eat and still overlook his work. . . . After this he is incessantly engaged—not a single minute of rest or relaxation being allowed him. At twelve o'clock, the engine stops, and an hour is given for dinner. The hands leave the mill, and seek their homes, where this meal is usually taken. It consists of potatoes boiled, very often eaten alone; sometimes with a little bacon, and [rarely] with a portion of [meat]. . . . As soon as this is effected, the family is again scattered. . . . Again they are closely immured from one o'clock till eight or nine, with the exception of twenty minutes, this being allowed for tea. . . . During the whole of this long period they are actively and unremittingly engaged in a crowded room and an elevated temperature, so that, when finally dismissed for the day, they are exhausted equally in body and mind.[52]

The hellish conditions in many of the factory towns were also vividly described by Friedrich Engels in *Conditions of the Working Class in England:*

> Heaps of refuse, offal and sickening filth are everywhere interspread with pools of stagnant liquid. The atmosphere is polluted by the stench and is darkened by the thick smoke of a dozen factory chimneys. A horde of ragged women and children swarm about the streets and they are just as dirty as the pigs which wallow happily on the heaps of garbage and in the pools of filth. . . . The inhabitants live in dilapidated cottages, the windows of which are broken and patched with oilskin. The doors and door posts are broken and rotten. The creatures who inhabit these dwellings and even their dark, wet cellars, and who live confined amidst all this filth and foul air—which cannot be dissipated because of the surrounding lofty buildings—must surely have sunk to the lowest level of humanity.[53]

These appalling conditions, and the extremes of wealth and poverty that resulted, aroused a deep sense of injustice about capitalism and the way it was working. Socialism was an impassioned response to such a brutal economic system. At one end of the socialist rainbow, the ideas of the nineteenth-century reformers were every bit as utopian as Plato's philosopher king or the fantasy of a perfectly equilibrated market economy. For instance, Claude Henri de Saint-Simon condemned what he called "the Hand of Greed" and proposed a transformation of society into small communal organizations, with economic production and distribution to be carried out by a benevolent bureaucratic state: "Rule over men would be replaced by the administration of things," he declared in a catchphrase that was later adopted by Karl Marx. Likewise, Pierre-Joseph Proudhon, whose writings also influenced Marx, expressed great hostility toward private property (as the epigraph to this chapter suggests) and wrote a book-length rant on the subject. Among other things, he argued that land rents were "unearned" and therefore not justified. Proudhon also envisioned an egalitarian distribution of wealth among the workers.[54]

Robert Owen in England also dreamed of a society that was organized into small collectives, but his approach was more pragmatic and incremental. During his twenty-five-year tenure, from 1800 to 1825, as a part-owner and manager of a textile mill, Owen instituted a series of farsighted reforms in the nearby village of New Lanark, including better housing, more affordable food, free education, and free health care. Thanks in part to these innovations, Owen's mill proved to be a great commercial success, and it soon became a model for the "company towns" of the late nineteenth and early twentieth centuries. Owen in his later years also became a prolific pamphleteer and political agitator and is credited with catalyzing the worldwide cooperative movement that exists to this day.[55]

Overshadowing all of these and other early socialist writers, however, were Karl Marx and his longtime partner Friedrich Engels. If, as Marx outrageously proclaimed, "religion is the opiate of the masses," Marxism during the early twentieth century became the opiate of the intellectual classes. Marx's writings inspired idealistic communist "cells" on university campuses and in coffeehouses around the world.

Although Marx was openly scornful of other utopian social schemes, claiming they were moralistic and unscientific, his own "scientific socialism" differed mainly in the means by which, according to Marx and Engels, a communist society would be achieved. Otherwise their basic vision—which provided only a few generalities—was very similar to those of other utopian socialists. Marx and Engels imagined an egalitarian society composed of many small voluntary communes, along with common ownership of "the means of production."

What sharply differentiated Marx and Engels from other utopians was their claim that this outcome was foreordained by an overarching economic and social dynamic that determined the ultimate course of human history. As Marx put it in the preface to his masterwork *Das Kapital,* communism would be achieved via "tendencies which work themselves out with iron necessity towards an inevitable goal."[56] These iron tendencies referred to Marx's theory (inspired by Hegel) that history is inexorably driven by a "dialectical" process consisting of social class conflict, revolution, and societal transformation. In the present "bourgeois" stage, Marx asserted, there is an implacable "class struggle" going on between capitalists and the working class, and this irreconcilable conflict is destined to culminate in the overthrow of capitalism and its replacement by a communist society.

Needless to say, this is only a bare-bones outline of a complex, sophisticated and deeply researched scholarly thesis that played a major role in the history of the twentieth century. It is worth noting, though, that Marx's views also evolved over time. For instance, the enemies of Marxism—who are legion—often ignore the fact that in their later years Marx and Engels conceded that the dialectic might be only a general tendency and that a communist society might also be achieved by peaceful "evolution." They considered England the most likely candidate for such a benign transformation.

By the same token, many people who called themselves Marxists freely interpreted his writings for their own purposes. On one famous occasion, Marx himself was moved to exclaim, "If this is Marxism, then I am not a Marxist." Indeed, had Marx and Engels lived long enough they would probably have been appalled by the brutal totalitarian regimes that arose in their names in the wake of the Russian and Chinese revolutions. Marx would likely have argued that the conditions for a dialectical transformation were not yet ripe in these preindustrial states. It is also worth noting that, in the generations since Marx and Engels died, an army of scholars have devoted their lives to interpreting, modifying, and attacking Marx's theoretical edifice.

Much more could be said about Marxism, but one other aspect of Marx's argument is especially relevant here—his theory of surplus value. This is what provided the moral foundation for his claim that the working class was being irretrievably exploited. Echoing John Locke's theory of property (mentioned in chapter 4) as well as the writings of Proudhon, Marx claimed that it is labor, after all, that is primarily responsible for the production of material goods and services. So, to simplify a bit, if the workers are able to produce "surpluses" (profits) beyond what is required for their subsistence, and if the property owners appropriate these excess profits for themselves, they are in effect stealing from the workers.[57]

Modern economists reject Marx's argument, of course. Many factors contribute to the production of profits in a complex business enterprise. Investment capital and entrepreneurship are extremely important ingredients, and so are technology, organization, management, and, needless to say, the division of labor (or, as I prefer to call it, the combination of labor).

The "value" of any product or service is therefore not a simple function of the labor involved. (Recall my use of the term *corporate goods.*)

Marx's argument is untenable, but so is the capitalist claim that a capital investment necessarily confers exclusive ownership and property rights and therefore entitles the investor to whatever surpluses are produced by the enterprise, or the "firm." It is certainly true that this is the way the capitalist system very often works, but this is not always the case and it is certainly not a reflection of some natural law. (I will return to this important issue later on.)

These days there remain very few utopian socialists or unabashed Marxists. However, at the other end of the socialist rainbow there have been many moderate, democratic reformers whose influence over the politics of the twentieth century was immense. Perhaps the most illustrious place in the social democratic firmament belongs to the Fabian Society in England, a multigenerational network of theorists and political activists that was founded in the late nineteenth century and has continued to influence British (and American) politics down to the present day. Over the decades, the Fabian Society's membership has included many of England's leading intellectuals and politicians (George Bernard Shaw, H. G. Wells, Virginia Woolf, Bertrand Russell, Graham Wallas, Havelock Ellis, Harold Laski, G. D. H. Cole, Clement Atlee, Ramsay MacDonald, Harold Wilson, Tony Blair, Gordon Brown, and many others), as well as many distinguished British Commonwealth leaders (modern India's founding father, Jawaharlal Nehru, Pakistan's Muhammad Ali Jinnah, Singapore's Lee Kuan Yew, and others).[58]

It is generally acknowledged, though, that the basic character of the Fabian Society was strongly influenced by two of its early leaders, Sidney and Beatrice Webb, whose prolific writings and political activism shaped the socialist agenda in England for the generations that have followed. For instance, it was the Webbs, and the early Fabians, who introduced the idea of a minimum wage and social insurance, including universal health insurance (though these ideas actually originated in Otto von Bismarck's Germany, as I noted earlier). The Fabians also promoted universal public education, slum clearance, and public housing long before these programs became pillars of the modern welfare state.[59]

However, the most radical and politically divisive of the early Fabian proposals called for nationalizing private lands and major industries in "the common interest." Nationalization of landholdings never came about in England, although landowners were heavily taxed during World War II and in the Labour (socialist) era that followed the war. On the other hand, the postwar Labour government did nationalize various industries (such as mining and the railroads), with mixed results overall. Eventually there was a political backlash, which resulted in some reprivatization and a resurgence of free market capitalism in the 1970s and 1980s under Prime Minister Margaret Thatcher—a course reversal that also occurred in our own country under President Ronald Reagan.

Elsewhere in Europe, various socialist and social democratic political parties have produced a diverse array of welfare states since the 1930s, with varying degrees of public and private ownership and a variety of social welfare programs. Yet, strangely enough, it is Sweden, one of the smaller European states, that has become a vortex of controversy between free market capitalists and promoters of the welfare state. The obvious reason is that Sweden offers the most generous package of social benefits of any developed nation, from free education and free health care to unemployment benefits that (depending on the circumstances) can run as high as 80 percent of a worker's salary for as long as five years! Sweden also boasts such amenities as one of the world's finest public transportation systems.

Of course, Sweden's generosity comes at a price. It has the second highest overall tax burden in the industrial world (behind only Denmark) according to the Organisation for Economic Co-operation and Development (OECD)—currently about 47 percent of GDP compared with about 27 percent in the United States. (Our tax burden is the lowest, believe it or not.)[60] But look at the results. According to data from the Central Intelligence Agency (CIA) *World Factbook* for 2009, comparing the percentage of the population in different countries that are living in poverty, Sweden is listed as having the world's third lowest at 6.5 percent, behind only Finland and Norway. (The poverty rate among Sweden's elderly population is only 1 percent, compared with a shocking 23 percent in the United States.) Sweden's infant mortality rate, a sensitive health indicator, is the second lowest in the world (after Singapore) at 2.75 per thousand live births, compared with 6.26 in the United States. Sweden also ranks tenth in life expectancy worldwide. (The United States, again, ranks fiftieth.)[61]

True, Sweden rates a somewhat less impressive twenty-third in GDP per capita at $38,200 (in 2009), compared with $47,500 in the United States. But Sweden's income is much more equally distributed, and the dollar figure does not factor in its lower overall cost of living and its many social benefits. For instance, the so-called Gini coefficient of personal income disparities between the top and bottom 20 percent of a country's population for 2002 placed Sweden at a low index number of 25 compared with 40.8 in the United States. (By 2008, our coefficient had climbed to 45, according to the CIA, well into what is considered the danger zone for social unrest.) In Sweden, the bottom 20 percent of the population received 9.6 percent of the income during that year while in the United States the figure was 5.2 percent.[62]

No wonder, then, that free market conservatives become apoplectic about Sweden. Some of the diatribes that can be found on the Internet these days make the place sound like a concentration camp. Of course, the bemused Swedes don't see it that way. Back in the 1970s and 1980s, conservative critics did have a case that the burden of the welfare state and a generally negative public attitude toward entrepreneurship and private enterprise caused stagnation in the Swedish economy, which ultimately resulted in a debt crisis. But what the Sweden-bashers overlook, or choose to ignore, is that this is now very old news. Sweden's economic slump was radically reversed in the 1990s, and today its economy is one of the most dynamic and competitive in Europe.

Even *Forbes* magazine, that self-described "capitalist tool," praised Sweden's impressive economic revival a few years back. Before the recent crisis dented Sweden's economy, GDP growth was running as high as 4 percent annually, with low inflation, strong productivity gains, low unemployment, robust wage increases, government budget surpluses, relatively low taxes on corporations, and a strong influx of investment capital. Over a six-year period, Sweden had experienced a 25 percent increase in the formation of new businesses. (Contrary to the mythology about Sweden, only about one-quarter of its business sector is publicly owned, and Sweden is currently working to reduce that percentage even further.)[63]

So, as the saying goes, What's not to like? Anyone in our country who has visited an IKEA store can plainly see that Sweden is quite capable of innovation, entrepreneurship, high-quality manufacturing, excellence in management and marketing, and highly competitive pricing. (A visit to Sweden itself can also be an eye-opener.) In short, Sweden has demonstrated that it is possible, after all, to have your capitalist cake and eat it too. Call it capitalism with a social conscience.

Nevertheless, "European socialism" remains a visceral pejorative among free market conservatives and the Republicans in Congress these days. But European mixed economies can hardly be called "socialism" in the term's original meaning—a truly egalitarian cooperative society with common ownership of the means of production and an equal sharing of wealth. Rather, they are capitalist states that are also committed to providing generous social welfare programs.

It's certainly true that social democratic regimes can become paternalistic, overregulated, bureaucratic, and all too often preoccupied with gaining and holding political power. Worse yet, they can be slow to adapt to change and can become obstacles to the innovation and entrepreneurship that are imperatives for economic survival in an increasingly competitive global economy. But Sweden has shown that these ossifying tendencies can also be resisted and reversed without abandoning the welfare state.

(I'll have more to say about the European welfare state model in the next chapter.)

By contrast, the form of schizoid state socialism that was tried out unsuccessfully, even disastrously, in the Soviet Union and Red China during the twentieth century was far more dysfunctional than the worst of the European socialist regimes. Soviet and Chinese communism represented a unique hybrid—a highly centralized, authoritarian, oppressive (and elitist) "planned economy," with the means of production controlled by the "state" and not the workers, and with a patchwork of subsidized employment programs and social services that were often inefficiently administered.

What economists disdainfully characterize as a "command economy" can actually be made to work under exceptional circumstances. During World War II, the United States government was highly authoritarian, even down to censoring mail, and controlled every aspect of our economy, but it was also highly successful because there was trust in a competent and honest government and broad public support for the war effort. It's not a formula for these times, unfortunately. (Of course, the Chinese have been pioneering a new economic model that might be called "authoritarian capitalism.")

One of the many problems with classical communitarian socialism, from Rousseau to Marx and beyond, is that it discounts the fundamental role of capital (economic surpluses) and entrepreneurship (new ideas, new technologies, risk taking, and dedicated hard work) as engines of economic development and change. Progress is not the ineluctable result of some inherent historical dynamic—as Marx and even some modern theorists seem to suppose. If you discourage initiative and enterprise, you are eating your seed corn (to use a venerable farm metaphor).

Surprisingly enough, many utopian socialists seemed cavalier about the importance of the division of labor and its role in our emergence and progress as a species. Adam Smith himself provided a classic example in *The Wealth of Nations.* At a pin factory that Smith had visited, ten workers performing ten different tasks were able to manufacture about forty-eight thousand pins a day. But if each of the laborers were to work alone, each one attempting to perform all the tasks associated with making pins rather than working cooperatively, Smith doubted that on any given day they would be able to produce even a single pin per man.[64]

Another way of looking at the pin factory, however, is in terms of how various specialized skills, tools, and production operations were *combined* into an organized system. It really should be called a "combination of labor." The system included not only the roles played by each of the workers, which had to be precisely coordinated, but also the appropriate machinery, energy to run the machinery, sources of raw materials, a supporting transportation system, and, not least, markets where the pins could be sold to recover production costs.

Not only that, but the pin factory required management—one or more persons responsible for hiring and training workers, for planning, for production decisions, for marketing and selling the pins, for payroll and bookkeeping, and so forth. And, of course, a capital investment was required to set up the factory in the first place. In other words, the economic benefits realized by the pin factory were the result of the total system, including a complex network of production tasks and cooperative relationships. Utopian socialists often seem oblivious to this organizational imperative—and so do some utopian capitalists, for that matter.

One of the most serious deficiencies in socialist theory, though, is its radical egalitarianism and extreme indifference to merit—our very real and significant differences in talent, experience, innovativeness, hard work, and personal achievement. As I noted earlier, it is just as unfair (inequitable) to ignore differences in our performance and our contributions as it is to indulge cheaters and free riders—concepts that are incomprehensible to utopian socialists. Indeed, the term merit is not even a part of the classical socialist lexicon.

This blind spot is also reflected in the utopians' hostility to private property as a desirable reward for merit. Private property is not simply a capitalist invention. Recall anthropologist Donald Brown's study, mentioned in chapter 4, in which he found that personal property is a cultural universal.[65] It's even found, to a degree, in some animals, most notably chimpanzees. Indeed, property rights are readily understood even by small children. Moreover, it's a legally sanctioned and widely accepted cultural institution that is thousands of years old. So the utopian socialists' lust for seizing and "socializing" private property on behalf of the people, or the workers (which occurred in the Soviet Union and Red China), is profoundly unfair. What may seem to some people to be well-meaning social amelioration is likely to be perceived by others as confiscatory—legalized theft.

On the other hand, the capitalists' claim to unrestricted property rights is also deeply unfair, especially when a (metaphorical) "pin factory" with many stakeholders is producing the wealth and when some of the stakeholders are experiencing "harm" in terms of their ability to meet their basic needs. When the conservative ideologue Grover Norquist was asked in an interview about taxing the wealthy to help the needy, he replied: "To do that, you would have to steal from the people who earned it and give it to people who didn't. And then you make the state into a thief."[66]

Of course, this assumes that the existing distribution of wealth is earned (merited), but this is often enough not the case. Some obvious examples are inherited wealth, the exorbitant CEOs' salaries and perks, the gluttonous bankers' bonuses, the politically sacrosanct farm subsidies, "coupon clipping" by unproductive shareholders who own "cash cow" stocks, and old-fashioned exploitation of workers in modern-day sweatshops. Various writers have also noted that the accumulation of wealth may simply be a matter of luck. The classic example is the hardscrabble Texas farmers in the 1920s and 1930s who happened to own farmland that sat on top of an underground pool of oil. For that matter, consider the fortuitous locations of the oil-rich Middle East countries.

By the same token, many unmet needs arise from the fact that people cannot find work in our capitalist economy or are incapable of doing so (like the aged, the sick, and young children). As I noted earlier, mutual aid for those who are legitimately in distress is a deeply embedded part of our heritage as a species and is reflected in the Golden Rule. So this impulse is also a part of our human nature, and the Grover Norquists are obvious outliers. (Norquist is also infamous for his desire to shrink government down to the size where he could "drag it into the bathroom and drown it in the bathtub.")

Related to the socialist attitude toward property is a congenital naiveté among the utopians about the role of competition in human societies. All the evidence we have about human nature indicates that reordering society without regard to the competitive aspect of our evolutionary heritage is biologically unsound. Our egoistic, competitive impulses are hardly confined to capitalist societies and are certainly not a capitalist invention. Indeed, over the centuries there have been a great many failed attempts to establish and sustain small cooperative communities, perhaps in response to some deep urge to reincarnate the tribal societies of our ancient past. All these experiments have failed, save a few exceptions that don't disprove the rule.[67]

But more important, competition can also play a vital positive role in a complex, dynamic economy. Generations of entrepreneurs bear witness that competition can energize the process of trial and error that is a necessary element of social and economic innovation. Competition has also traditionally served as a safeguard against the abuses that can arise from holding a monopoly. It's rather like drinking wine. Consumed in moderation, competition may have some beneficial effects. But overdoing it can have all manner of destructive consequences.

It is also worth reminding ourselves that one of Darwin's most important insights had to do with the role competition plays in evolution. Competition, sometimes direct but more often an indirect result of the challenges associated with earning a living in nature, has been an engine of evolutionary change for billions of years. Differential survival and reproductive success constitute the very essence of Darwin's concept of natural selection. However, in human societies competition is a very volatile form of social behavior that can easily turn into mutually destructive "warfare," where anything goes. To be constructive, it must be strictly contained and aggressively refereed. As Aristotle warned, humankind needs a "social cage."

Rewards for merit play an especially important role in socially approved forms of competition, just as punishments are usually prescribed for not "playing by the rules." Consider this hypothetical situation: In the swimming competition at the 2008 summer Olympic Games, would it have been fair to award only "participation medals" to each of the athletes, as is often done in preschool sporting events? In that case, Michael Phelps would have returned home with only a participation medal to display for his record-breaking accomplishments (eight gold medals). Come to think of it, in the absence of any personal recognition for his competitive achievements, Phelps might not have been motivated to try quite so hard.

A final defect in utopian socialist theory is that it typically omits any reference to "reciprocity"—duties and obligations that are commensurate with the benefits received. Thus Marx famously proclaimed that communism would be based on the principle of giving "to each according to his need." So far, so good. But Marx coupled this with a call to take "from each according to his ability." This vague formulation can be interpreted in different ways, some of which might be very inequitable. For example, if one farmer works harder and more skillfully and grows an abundant surplus of wheat, far beyond the yields of an incompetent, lazy neighbor, should that surplus be appropriated by the community simply because he or she has more? Or should there be some attention to equity—to merit?[68]

In sum, the basic problem with utopian socialism is that it's utopian. It's not grounded in a realistic understanding of human nature, nor does it have any real appreciation for the challenges associated with organizing and sustaining the collective survival enterprise—much less adapting it to changing circumstances. In its own way, it is also profoundly unfair.

What can we conclude? First of all, capitalism both in theory and in practice is very often myopic and sometimes in denial about the basic needs that constitute biological imperatives for every one of us. In this fundamental respect, we are all unequivocally "equal," and any economic system that tolerates great disparities of power and wealth to the point that it causes serious "harm" to other members of a society is deeply unfair. Not even a self-respecting libertarian can justify this.

On the other hand, the egalitarian ideal of utopian socialism is also deeply unfair. Radical socialists often ignore the principle of "proportionate equality," or rewards for merit. Indeed, socialists often treat equality and equity as synonyms and fail to differentiate between procedural and substantive equality.

But if socialists are wrong about the claims of merit, capitalist economists are profoundly wrong in treating labor as an input that should be used as efficiently as possible. Many studies, and many students of business management practices, have amply documented that the best-run firms, by and large, are the ones that treat the workers as "stakeholders" in the organization (a concept I will return to in the final chapter). Longtime business writer Andrea Gabor, in her insightful historical study *The Capitalist Philosophers,* put it this way:

> The long-dominant metaphor for the corporation—that of a machine—is a root of the malaise. To think of an organization as a mechanical device is to discount the value of human creativity and the possibility that organizations can foster a sense of purpose in an almost organic sense. People are the source of competitive advantage. . . . The conclusions [of a Shell Oil Company management study] put a premium on trust, civic behavior, the development of individual potential, and leadership as stewardship. In other words, they mark a resounding endorsement of the stakeholder-versus-shareholder-dominated philosophy of management.[69]

Both free market capitalism and utopian socialism can also be accused of having unrealistic attitudes toward the role of government. Radical socialists seem intent on minimizing or eliminating government in the naive belief that a socialist society would be self-regulating. On the other hand, radical capitalists like the Ayn Rands and Grover Norquists see the state as a monolith that mainly serves as an obstruction to private enterprise. They devoutly believe that the private sector can provide just about any social service more "efficiently." But the fact is that sometimes government can do the job better. Consider the case of health insurance. The administrative costs for the highly popular Medicare program run about 3 to 5 percent, whereas the costs for private, for-profit health insurance plans typically run between 25 and 30 percent (including profits and, of course, those gargantuan advertising and lobbying expenses).[70]

Finally, both utopian capitalists and utopian socialists have simplistic, one-dimensional views of human nature—stereotypes that have inspired false assumptions and fatally destructive policies. Neither the rational, calculating, egoistic *Homo economicus* nor the cooperative, caring, altruistic "socialist man" can encompass the dualities and the diversity of humankind. All the evidence suggests that a more nuanced and complex view of human nature is essential. So it's time to move on—to redefine the social contract in a way that is at once more realistic and more fair-minded. In fact, our current economic crisis demands it. We can no longer afford the luxury of what Marx called a "false consciousness." As the French intellectual André Gide is reputed to have observed, "The color of truth is gray."

7 * *Toward a Biosocial Contract*

> The first goal of justice is to create a *modus* vivendi so that life can go on, not only in the next few minutes, but also indefinitely into the future.
>
> GARRETT HARDIN

> Competition has been shown to be useful up to a certain point and no further, but cooperation, which is the thing we must strive for today, begins where competition leaves off.
>
> FRANKLIN D. ROOSEVELT

If the many fans of Adam Smith would take the trouble to read *The Wealth of Nations,* they would find that Smith harbored all manner of subversive thoughts. (Remember his observation, quoted earlier, about the conspiracy among the "masters" to hold down the wages of laborers, and his support for the Golden Rule?) Sometimes, in fact, Adam Smith reads like Karl Marx—or rather, Marx reads like him, since he predated Marx by many years. Here is another example: "Civil government, so far as it is instituted for the security of property, is in reality instituted for the defense of the rich against the poor, or of those who have some property against those who have none at all."[1]

Like other early liberal economists, Adam Smith viewed the English government of the late eighteenth century as a bastion of privilege that unfailingly supported England's hereditary, landowning aristocracy. Smith championed free market capitalism because, among other things, he saw it as a means for abolishing the entrenched social classes and the extremes of wealth and poverty. Free markets, he believed, would break the stranglehold of the landed aristocracy and spread new wealth.

Of course, things didn't work out quite the way Smith had expected. A rising new propertied class—wealthy capitalists—soon gained control of the levers of government and invented new forms of poverty. Recall that it took several generations to reform the inhumane factory working conditions that arose during the Industrial Revolution. Karl Marx was a personal witness to the Great Transformation (the title of political economist Karl Polanyi's classic study of the subject) and contemptuously characterized the nineteenth-century English state as a "handmaiden" of capitalism.[2] It's not surprising, therefore, that Marx viewed free market capitalism as the problem, not the solution (to turn Ronald Reagan on his head) and proposed a remedy that was radically opposed to the invisible hand. One of Marx's detractors dubbed his revolutionary vision "apocalyptic socialism."

In retrospect, it's clear that both Smith and Marx—and a great many of their soul mates and camp followers—underrated the importance of political power in determining "who gets what, when, how"—to quote the subtitle of a famous tract by twentieth-century political scientist Harold Lasswell.[3] To repeat Plato's observation in *The Republic,* perhaps the most fundamental political challenge of all is who will guard the guardians. ("Quis custodiet ipsos custodes?") Whose interests are shaping the rules of the game? Is the state an impartial arbiter that ensures a degree of fairness among the various stakeholders? Is it able to achieve a balance between the three fairness criteria discussed earlier—equality (basic needs), equity (merit), and reciprocity? Or is it the captive of one class of stakeholders at the expense of the others—a predator state? Or worse, is it the instrument of some dictator's personal agenda?

A starting point for addressing these questions is the masterful study of the modern welfare state by the sociologist Gøsta Esping-Andersen, *The Three Worlds of Welfare Capitalism.* As Esping-Andersen explains, the rise of the welfare state has been closely tied to the progressive democratization of Western political systems over the past two centuries through the expansion of the voting franchise and the political mobilization of middle-class and working-class citizens. "Politics not only matters but is decisive," Esping-Andersen concludes, endorsing the thesis originally put forward by Karl Polanyi in his famous study.[4] Just as Plato and Aristotle had prescribed, it seems that a mixed government where all the various stakeholders are empowered and represented can do a better job of avoiding the extremes of wealth and poverty and realizing a modicum of social justice. Once again, Lord Acton got to the heart of the matter: "The danger is not that a particular class is unfit to govern. Every class is unfit to govern."[5]

To achieve the kind of modus vivendi that Garrett Hardin was referring to in the epigraph above, there must be a broad political consensus that social justice is a core social value; otherwise political compromises may be impossible to achieve. The great ideological war between capitalism and socialism remains unresolved in part because ideologues on both sides continue to adhere to sharply opposing models of how a society should be organized and governed—and how the pie should be divided. Consider, for instance, the contrast between Michael Moore's sardonic movie *Capitalism: A Love Story* and the shrill and angry rants against liberals by Glenn Beck and others on any given day at Fox News.

It's no surprise, therefore, that there has been unrelenting trench warfare in American politics during the past thirty years over such issues as taxes, government regulation, social insurance, and the role of government in health care, education, and other social programs. There has been a strong tilt in this country (and in the United Kingdom as well) toward the free market capitalist model. On the other hand, some continental European countries have been able to maintain a better balance among the various political constituencies. Even as we have witnessed some unraveling of the safety net in the United States during the past three decades, further improvements have occurred in the social welfare systems of countries like Denmark, the Netherlands, and Sweden. Nevertheless, even in these societies there remains a deep ideological fault line. As Esping-Andersen explains, "To a social democrat, reliance on the market for the basic means of welfare is problematic because it fails to provide inalienable rights and because it is inequitable. To a *laissez-faire* [conservative], reliance on the welfare state is dangerous because it cripples freedom and efficiency [and redistributes wealth, of course]."[6]

While many conservatives still self-righteously cling to the free market model, many social democrats remain devoted to the egalitarian ideal and are frustrated by the persistence of social stratification—the sharp differences in wealth and privilege that capitalism can produce. Social democrats still want to "emancipate" the middle and working classes from dependence on "the cash nexus" and the vicissitudes of the marketplace (especially such vicissitudes as our current recession).

To this end, some European social democrats have embraced the concept of "social rights," which they believe should be given equal status with property rights. They speak of "de-commodification"—meaning that social services should be provided as a right and not as a commodity that must be purchased in the marketplace (such as our predominately private health insurance system in this country). They dream of a time when, in Esping-Andersen's words, people will have "a genuine option to working" and when "a person can maintain a livelihood without reliance on the market." Of course, this objective is highly objectionable to capitalists, and to many others as well, because it sounds like a license to become free riders who are not obligated to contribute a fair share to the economy that supports them. In other words, the social rights ideology seems a bit insensitive about the principle of reciprocity.

One effort to bridge this ideological divide has been called the "Third Way"—a distinctively Anglo-Saxon approach that some historians trace back to the Fabians, to FDR's New Deal, to Britain's Beveridge Report at the end of World War II, and to British prime minister Harold Macmillan's "Middle Way" conservatism in the 1950s. However, the term Third Way and its contemporary elaboration have been credited to the British sociologist Anthony Giddens, who characterizes his approach as a "centrist" political strategy designed to achieve a balance between free market capitalism and the welfare state. In Giddens's words, the Third Way has an "unswerving belief in the merits of the free market. . . . The Third Way is in favour of growth, entrepreneurship, [free] enterprise and wealth creation. But it is also in favour of greater social justice and sees the state playing a major role in bringing this about."[7] Prominent supporters of the Third Way have included prime ministers Tony Blair and Gordon Brown in the United Kingdom and President Bill Clinton in the United States, as well as political leaders in Canada, Australia, New Zealand, and a few other countries.

Although the Third Way has been described by some of its supporters as a "radical middle" or a "radical center," it is in fact an old-fashioned political compromise—a calculated effort to placate corporate business interests in return for being able to make some incremental improvements in the existing welfare state. It has, in fact, allowed for some tinkering with the machinery, but it has certainly not achieved any serious social re-engineering. In practice, the Third Way has been permissive toward the abuses of free market capitalism, indulgent toward the rapidly increasing gap between the rich and the poor, and unable to make significant progress in patching up an increasingly tattered social safety net in this country. Indeed, by endorsing and facilitating financial "deregulation" in the 1990s, Third Way supporters in this country contributed to the financial bubble and its tragic consequences.

If there was a moment in history when the Third Way crashed and burned (at least in the United States), it was in 1993, when President Clinton's health care reform initiative, led by Hillary Rodham Clinton, collapsed in the face of fierce opposition from private industry and conservatives. For the next decade and more, a strident, self-righteous, even angry conservatism—epitomized by House Speaker Newt Gingrich's "Contract With America" (mocked by liberals as a "contract on America")—dominated American politics. Even President George W. Bush's "compassionate conservatism" mantra during the 2000 presidential election campaign proved to be a case of deceptive packaging and was soon dropped in favor of tax cuts for the wealthy, restrictions on family planning and stem-cell research, a hard-nosed "No Child Left Behind" policy, a plan to privatize Social Security, and the like. (For another Third Way vision, see endnote 7.)

The task that now lies before us, therefore, is to move beyond the clichés about capitalism and socialism and reboot the public philosophy with a better grounded vision of human nature and the underlying purpose of a human society. If every crisis is also an opportunity, as the old saying goes, it is above all an opportunity to rethink the game plan we have been following and, more important, to decide which game we want to be playing.

In fact, an organized, complex society is more than a market and more, even, than a "civilization." To reiterate, it is a purposeful, biologically based collective survival enterprise. All of us are stakeholders in this enterprise, and we are all dependent on it in various ways. An appreciation of this deep interdependency was articulated by, of all people, Adam Smith in this classic passage from *The Wealth of Nations:*

> In civilized society [man] stands at all times in need of the cooperation of and assistance great multitudes. . . . [M]an has almost constant occasion for the help of his brethren. . . . Observe the accommodation of the most common artificer or day-labourer in a civilized and thriving country, and you will perceive that the number of people of whose industry a part, though but a small part, has been employed in procuring him this accommodation, exceeds all computation. The woolen coat, for example, which covers the day-labourer, as coarse and rough as it may appear, is the product of the joint labour of a great multitude of workmen [shepherds, wool sorters, combers, dyers, spinners, weavers, fullers, dressers, merchants, transporters, shipbuilders, sailors, sailmakers, rope makers, toolmakers, miners, furnace builders, bricklayers, lumbermen, charcoal makers, millwrights, forgers, blacksmiths, and more]. . . . If we examine, I say, all of these things, and consider what a variety of labour is employed about each of them, we shall be sensible that without the assistance and cooperation of many thousands, the very meanest person in a civilized country could not be [provisioned].[8]

Every complex society consists not only of individuals and families but of a multitude of pin factories (purposeful social organizations), ranging from your favorite local restaurant to multinational corporations with hundreds of thousands of employees and, not least, the plethora of government agencies and institutions. It is a vast, organized combination of labor where cooperation predominates (and must prevail, after all), where competition is largely unavoidable, often useful, and sometimes destructive, and where conflicts over our separate interests are ubiquitous. Yet we remain deeply interdependent. So the challenge for every society with a commitment to sustainability for the long run is to mediate among these conflicting interests and, as Garrett Hardin urged, find the modus vivendi that achieves social justice. It is time, therefore, to rewrite the social contract.

We met up with the social contract in chapters 3 and 4, where I noted that the concept has traditionally been employed by philosophers as a theoretical gambit—a device for reasoning about the nature and purpose of a society and the role of the state. Although the basic idea traces back to

Plato and other ancient Greek philosophers like Epicurus, it is commonly associated with the so-called social contract theorists of the seventeenth and eighteenth centuries—such as Rousseau, Hobbes, and Locke—and, more recently, John Rawls. In chapter 4, I described Rousseau's fantasy about free individuals voluntarily forming egalitarian communities in which everyone has an equal voice and all are equally subject to the collective (general) will. I also noted how Thomas Hobbes, in contrast, envisioned a natural condition of anarchic violence. He proposed that, for mutual self-preservation, our ancestors conceded their individual rights to the sovereign will of the state. John Locke, on the other hand, conjured a more benign state of nature in which free individuals voluntarily formed a limited compact for their mutual advantage but also retained various residual rights.

The philosopher David Hume and many others since have made hash of this line of reasoning. In a devastating critique, *A Treatise of Human Nature* (published in 1739–40), Hume rejected the claim that some deep property of the natural world—"natural laws"—can be used to justify our moral precepts. Hume specifically undercut the reasoning of both utilitarians like Bentham, with his claims for the priority of the pain/pleasure principle, and the social contract theorists, with their claims about our rights and duties. Among other things, Hume pointed out that even if the origins of human societies actually conformed to such hypothetical motivations and scenarios—which we now know they did not—we have no logical obligation to accept an outdated social contract that was entered into by some remote ancestor.[9] With the demise of the natural law argument, social contract theory has generally fallen into disfavor among philosophers, with the important exception of the work of John Rawls. His 1971 book *A Theory of Justice* represents a landmark in social theory.[10]

Whatever its flaws, Rawls's formulation of an "original position" (a hypothetical analogue of the state of nature) provoked a widespread reconsideration of what constitutes fairness and social justice and, equally important, what precepts would produce a just society. As I noted in chapter 2, Rawls proposed two complementary principles: equality in the enjoyment of freedom (a concept fraught with complications) and affirmative action, in effect, for "the least advantaged" among us. This would be achieved by ensuring that the poor have equal opportunities and that they receive a relatively larger share of any new wealth whenever the economic pie grows larger. Although Rawls's work has been exhaustively debated by philosophers and others over the past three decades, it seems to have had no discernible effect outside the academic echo chamber, considering our deplorable economic statistics and the negative trends summarized at the beginning of chapter 6.

However, there is one other major exception to the general decline of social contract theory that is perhaps more significant. Over the past two decades, a number of behavioral economists, game theorists, evolutionary psychologists, and others have breathed new life into this venerable idea with a combination of rigorous, mathematically based game theory models and empirical research. I discussed some of the contributions of game theory to the fairness debate in chapter 4, along with the work of Cosmides and Tooby on social exchange and reciprocity. Some theorists, most notably the mathematician turned economist Ken Binmore and philosopher Brian Skyrms, have taken this work an important step further and have sought to use game theory as a tool for resuscitating social contract theory on a new footing.

Binmore's work, culminating in his 2005 book *Natural Justice*, is especially convergent with what I will be proposing here, although it also diverges in some important respects. Binmore describes his approach as a "scientific theory of justice" because it is based on an evolutionary/adaptive perspective as well as the growing body of research in behavioral and experimental economics regarding our evolved sense of fairness plus some powerful new insights from game theory.

Briefly, Binmore defines a social contract in very broad terms as any stable "coordination" of social behavior—like our conventions about which side of the road we drive on or pedestrian traffic patterns on sidewalks. Any sustained social interaction in what Binmore refers to as "the game of life"—say a marriage, a car pool, or a bowling league—represents a tacit social contract if it is stable, efficient, and fair. (He appreciates that the human psyche does indeed include a "deep structure of fairness.")

To achieve a stable social contract, Binmore argues, a social relationship should strive for an equilibrium condition—an approximation of a Nash equilibrium in game theory. The rewards or "payoffs" for each of the players should be optimized so that no one can improve on his or her own situation without exacting a destabilizing cost from the other cooperators (see chapter 4). Ideally, then, a social contract is self-enforcing. As Binmore explains, it needs no social "glue" to hold it together because everyone is a willing participant and nobody has a better alternative. Binmore also invokes the image of a masonry arch that requires no mortar (a metaphor that traces back to Hume).

The problem with this formulation—as Binmore recognizes—is that it omits the radioactive core of the problem: How do you define fairness in substantive terms? Binmore asserts that "rough reciprocity" must be a central guiding principle. He calls it the "mainspring" of fairness. But this is not so easy to measure, and the world is filled to overflowing with social conflicts where there are deep differences of opinion about what is fair, and where the game looks like a tooth-and-claw struggle with a zero-sum payoff matrix. Indeed, when the issue is how to divide a cake among various stakeholders, reciprocity may not offer a solution at all.[11] To reiterate my quotation from Binmore in chapter 4, game theory "has no substantive content. . . . It isn't our business to say what people ought to like." Binmore rejects the very notion that there can be any universals where fairness is concerned. "The idea of a need is particularly fuzzy," he tells us.[12]

In other words, Binmore's version of a social contract involves an idealization, much like Plato's republic, or utopian capitalism, or Marx's utopian socialism. It has no specific fairness content. Binmore claims that a social contract exists if there is a stable, efficient social relationship. As he says, it is merely a consensus about the rules of the game (the process). And if it is voluntarily accepted, it must also be fair by definition. Binmore is hardly insensitive to the social consequences of the game of life. In one passage he pointedly asks, "Who thinks it fair that every dollar should be regarded as equally well spent, whether allocated to a hungry child or a bloated [billionaire]?"[13] But in the end Binmore chooses to remain above the substantive fray and, in so doing, glosses over the fundamental issue.

As I noted in the earlier discussion, what looks like willing consent (what Binmore would call an equilibrium condition) may mask a vector sum of coercive forces—from physical threats to a lack of bargaining power (say, when you can be fired for not following orders), a lack of relevant information (as in Lilly Ledbetter's pay discrimination case), a lack of financial resources (money to hire a lawyer, for instance), a threat of punishment for "making waves" (being sidelined or passed over for promotion), and especially a lack of available alternatives (for example, when jobs are scarce).

Consider the many millions of exploited laborers throughout history who grudgingly accepted harsh, even punishing working conditions and starvation wages because they did not have a choice, given the imperative of meeting their basic needs. Most factory workers in those "dark satanic mills" of the Industrial Revolution had no realistic choice. Binmore acknowledges that many social equilibrium conditions are possible when coercion is involved. Can we really call such severely imbalanced relationships "social contracts"? I think not.

I prefer to stick more closely to the traditional meaning of a social contract as a truly voluntary bargain among various (empowered) stakeholders over how the benefits and obligations in a society are to be apportioned among the members. This also accords with the much-used colloquial meaning of the term. What I call a "biosocial contract" is distinctive in that it is grounded in our growing understanding of human nature and the basic purpose of a human society, as discussed in earlier chapters. It is focused on the content of fairness, and it encompasses a set of specific normative precepts. Plato and Aristotle are my touchstones, not John von Neumann or the game theorists.

In the game theory paradigm, the social contract is all about harmonizing our personal interactions. Well and good. But in a biosocial contract, the players include all the stakeholders in the political community and substantive fairness is the focus. The underlying purpose of a biosocial contract is captured in this much-quoted passage from the great eighteenth-century conservative philosopher Edmund Burke:

> Society is indeed a contract. . . . [But] the state ought not to be considered as nothing better than a partnership in trade . . . to be taken up for a little temporary interest, and to be dissolved by the fancy of the parties. As the ends of such a partnership cannot be obtained by many generations, it becomes a partnership not only between those who are living, but between those who are living, those who are dead, and those who are to be born. Each contract of each particular state is but a clause in the great primeval contract of eternal society.[14]

A biosocial contract is about the rights and duties of all the stakeholders in society, both among themselves and in relation to the "state." It is about defining what constitutes a "fair society." It is a normative theory, but it is built on an empirical foundation. I believe it is legitimate to do so in this case, because life itself has a built-in normative bias—a normative preference, so to speak. If we do, after all, want to survive and reproduce within the collective survival enterprise—if this is our shared biological objective—then certain principles of social intercourse follow as essential means to this end.

First and foremost, a biosocial contract requires a major shift in our social values. The deep purpose of a human society is not, after all, about achieving growth, or wealth, or material affluence, or power, or social equality, or even about the pursuit of happiness (although I wish it Godspeed). It is about how to further the purpose of the collective survival enterprise. It compels us to focus on meeting our shared survival and reproductive needs. It also requires us to give priority to the overriding importance of social cooperation without denying the contingent benefits of competition. However, it is also important to recognize differences in merit and to reward them accordingly. Finally, there must also be reciprocity, an unequivocal commitment from all of us to help support the survival enterprise, for no society can long exist on a diet of altruism. Altruism is a means to a larger end, not an end in itself, as some of our theologians and moral philosophers would have us believe. It is the emotional and normative basis of our safety net.

As I suggested in earlier chapters, a biosocial contract encompasses three distinct normative (and policy) precepts that must be bundled together and balanced in order to approximate the Platonic ideal of social justice. These precepts are as follows:

- *Goods and services must be distributed to each of us according to our basic needs (in this, there must be equality).*
- *Surpluses beyond the provisioning of our basic needs must be distributed according to "merit" (there must also be equity).*
- *In return, each of us is obligated to contribute proportionately to the collective survival enterprise in accordance with our ability (there must be reciprocity).*

The first of these precepts involves a collective obligation to provide for the common needs of all our people. To repeat the term I borrowed from *Star Trek,* this is our "prime directive." Although this precept may sound socialistic—an echo of Karl Marx's famous dictum—it is at once far more specific and more limited. It refers to the fourteen primary needs domains detailed in chapter 5. Our basic needs are not a vague, open-ended abstraction or a matter of personal preference. They constitute a concrete but limited agenda, with measurable indicators for assessing outcomes. It does not entail a wholesale redistribution of wealth.

Recall also that the Survival Indicators framework encompasses individual differences and context-dependent "instrumental needs" that are subject to change throughout the life cycle, as well as needs that are related to reproduction and the nurturing of dependent offspring. It should go without saying that both the marketplace and a variety of other forms of collective action (inclusive of government programs and policies) can play a role in meeting our basic needs. There is no one right way to accomplish this objective. In fact, the great majority of our population's needs are satisfied by the private sector and by voluntary action, and appropriately so. In short, this precept is not a mandate for overthrowing capitalism but rather a prescription for harnessing it more closely to our basic purpose as a society. (I will return to this issue in the last chapter.)

On the other hand, our fourteen basic needs domains are much broader than "needs" usually implies. As we saw earlier, our basic needs go well beyond soup lines, food stamps, homeless shelters, even health programs for the poor like Medicaid. Indeed, they provide a shopping list for a decent—if modest—living, including nurturing the next generation. It is about "no child left behind" in every sense of the term. And because survival and reproduction are universal biological imperatives, the moral claim on behalf of our basic needs takes priority.

In other words, our basic needs are not a matter of free choice. A failure to provide for these needs inevitably causes "harm." If there is "a right to life," as many conservatives claim, it does not end at birth: it extends throughout our lives. Life is prior to liberty, and prior to property, both in the litany of our constitutional ideology and as a matter of common sense. In fact, when forced to choose, most of us adhere to John Locke's rank ordering. Life comes first, and we cannot be free if our basic needs are not satisfied. We therefore have a sacred obligation to provide for the basic needs of all members of our society, and we are obviously falling far short of this goal. When Bill Moyers, in his interview with Bill Gates a few years back, asked the Microsoft founder why global incomes are so unequally distributed, Gates responded, "It's a mistake."

Should a moral claim on behalf of our basic needs be considered a natural right? Surprisingly enough, even the dour Thomas Hobbes seemed to think so. In another of his books, *The Elements of Law,* Hobbes argued that we retain a right to self-preservation, which includes "the use of fire, water, free air, and a place to live in . . . [and] all things necessary for life."[15] As many critics have pointed out, any claims for natural rights are, in reality, only social constructs—norms or codified laws that are socially accepted. So if we all agree to accept the principle that there is a mutual obligation to ensure that the basic needs of all our people must be satisfied and that we will do so collectively for those who are unable, for whatever reason, to provide for their own needs, then it can rightly be treated as a social rather than a natural right.

The idea that we have an unqualified social obligation to provide for the basic needs of all our "brethren" (in Adam Smith's term) is also embodied in the Golden Rule—the universal moral principle I discussed in chapter 2—and it has been reflected as well in the writings of such diverse modern figures as Franklin D. Roosevelt, Amartya Sen, John Rawls, Barack Obama, and many others. Indeed, if conservatives and libertarians are prepared to honor their own explicit disclaimers about not doing "harm," then they too must acknowledge a duty to support those who are seriously deprived. Anything less could be called negligent selfishness.

This conclusion is not as radical as it may seem. Numerous public opinion surveys over the years have consistently shown that people are far more willing to provide support for the genuinely needy than the Scrooges among us would lead one to believe. For instance, an ABC/*Washington Post* poll in the 1990s found twice as many people were willing to pay more taxes to reduce poverty as were opposed to the idea. Another poll by *Time* magazine found that three-quarters of the respondents agreed with the proposition that government should guarantee that everyone has enough to eat and basic shelter. More surprising still was a Gallup national survey in 1998 showing that respondents earning more than $150,000 a year overwhelmingly supported greater government assistance for the poor (67 percent), and 24 percent of those who expected they would do better economically in the next five years actually favored higher taxes on the rich. As Samuel Bowles and Herbert Gintis pointed out in an article a few years back titled "Is Equality Passé?" our evolved sense of fairness and empathy toward the misfortunes of others manifests itself in what the authors characterized as a "basic needs generosity."[16]

Even more compelling, I believe, are the results of a famous series of social experiments regarding distributive justice conducted by political scientists Norman Frohlich and Joe Oppenheimer and their colleagues, as detailed in their 1992 book *Choosing Justice*.

Frohlich and Oppenheimer set out to test whether ad hoc groups of "impartial" decision makers behind a Rawlsian "veil of ignorance" about their own personal stakes would be able to reach a consensus on how to distribute the income of a hypothetical society. The subjects, all college students, were asked to choose among four policy options: (1) John Rawls's principle of favoring the least advantaged; (2) economist John Harsanyi's principle of maximizing the average income per capita (in other words the utilitarian "greatest good" model); (3) maximizing the average income but allowing for a minimum "floor" of income for everyone; and (4) minimizing the gap between the richest and poorest citizens. The experiments were replicated (with some variations) eighty-one times in four locations: Florida, Maryland, Manitoba (Canada), and Poland. (Some additional experiments were excluded from the final tabulations owing to large differences in methodologies that were deemed to have biased their results.)

Frohlich and Oppenheimer found that the experimental groups consistently opted for policy number 3—striking a balance between maximizing income (providing incentives and rewards for "the fruits of one's labors," in the authors' words) and ensuring that there is an economic minimum for everyone (what they called a "floor constraint"). The overall results were stunning: 77.8 percent of the groups chose to assure a minimum income for basic needs, while 12 percent preferred the Harsanyi (utilitarian) option and only 1.2 percent supported Rawls's "difference principle"—enhancing the income of the least advantaged. The remaining 8.6 percent preferred to minimize the income gap.

To quote from Frohlich and Oppenheimer's conclusions: "The groups wanted an income floor to be guaranteed to the worst off... a safety net for all individuals. But after this constraint was set, they wished to pursue incentives so as to maximize production and average income.... The desire to find a compromise between incentives and security... seems predicated on a desire to balance claims of entitlements and needs, while preserving incentives for productivity." (The Frohlich and Oppenheimer experiments have since been replicated many times by researchers in Australia, the Philippines, Japan, and elsewhere in the United States.) [17]

The results of these important experiments also lend strong support to the second of our three fairness precepts, which squarely addresses the problem of equity. How can we also be fair-minded about rewarding our many individual differences in talents, performance, and achievements? The equity precept taps an important element of the deep psychology of fairness—our strong desire to be recognized and honored for our efforts and achievements. This recognition may or may not take the form of material rewards. Nonmonetary forms of acknowledgment—from audience applause to Boy Scout/Girl Scout merit badges and mass-produced, low-cost sports trophies—are often sufficient as rewards; social approval by itself is a powerful motivator of human behavior. But in the economic sphere, income is obviously of primary importance.

However, merit, like the term fairness itself, has an elusive quality; it does not denote some absolute standard. It is relational and context-specific, subject to all manner of cultural norms and practices. In general, it implies that the rewards a person receives should be proportionate to effort, or investment, or contribution. For example, no sane person would ever claim to have won the lottery on merit.

Indeed, in the economic sphere merit is not simply a matter of what the recipient thinks is fair treatment but reflects what is socially acceptable. As with fairness in general, merit very often has to split the difference. When you are asking others to reward you for your efforts, they are entitled to be stakeholders in the decision. Markets often do this in an impersonal way. If you are running a business, your sales volume is a direct measure of how others value your goods and services, though there can be many distortions and biases even in this supposedly objective measure.

Consider the example of Pierre Omidyar. In March 2008, he was declared by *Forbes* magazine to be the 120th richest person in the world, with a net worth of $7.7 billion. After the stock market meltdown, it was estimated that his wealth was only $3 billion—still a pretty comfortable nest egg—and there has been some recovery in the market since then. How did Omidyar achieve this exalted financial status? In 1995, at age twenty-eight, this Iranian American computer programmer and sometime entrepreneur came up with the idea for a local online auction site—now global in scope and known as eBay.

At first Omidyar offered his services free. Presumably the intangible rewards were sufficient for him. But as eBay grew rapidly in popularity, Omidyar began charging his users to cover his growing Internet server costs. Then, after three years of explosive growth, Omidyar was persuaded to take the company public with a stock offering that made him an instant billionaire. (He personally held 178 million shares.) By then he had already handed over the day-to-day management of the company to somebody else, and he has been devoting himself to various philanthropic ventures ever since.[18]

It could be argued that Omidyar's wealth is merited because this is the value the stock market places on eBay. The stockholders, after all, put up their money to show their tangible support. But the stock market is only one, often deceptive, measure of economic merit, not to mention social merit. (Remember the cigarette makers?) A company's stock price can be affected by many extraneous factors, including general economic or market conditions, "irrational exuberance" (Alan Greenspan's prophetic label for the high-tech bubble in the 1990s and the housing bubble of the past few years), the actions of market day traders (those unavoidable parasites who feed off market activity and are always trying to turn a quick profit), mutual fund managers who are bent on puffing up their portfolios but who will also sell out at the drop of a stock index, and so on.

The great virtue of capitalism, for all its faults, is that it encourages and rewards innovation, entrepreneurship, risk taking, and the "blood, toil, tears and sweat" (to borrow another bit of Churchillian rhetoric) that are required to create and run a successful business. It provides material incentives for behaviors that, ideally, serve the public good. At least this is the economics textbook version of the story.

However, as we have recently rediscovered, at great cost, a clear distinction must be drawn between, say, the excessive salaries and perks of corporate CEOs, or the billion-dollar bonuses of hedge fund managers, and any reasonable standard of economic merit. Such abuses are more meretricious than meritorious. Indeed, it severely deflates the value of merit when an outcome is unearned, or unfairly attained. Cheating on a final exam is one obvious example. Nobody, including the cheater, has any illusion that the resulting grade is a measure of merit.

The bottom-line justification for capitalism has always been that the wealth it creates is earned by the owners and is therefore deserved (merited). Moreover, it also supposedly benefits society (though it is not shared equally, of course). Often this is true, but often it is not. The wealth that is generated may be unearned by any reasonable definition of "merit" and, in the extreme, may actually cause "harm" to the rest of society. It could hardly be said that Bernie Madoff deserved the wealth he skimmed from his Ponzi scheme.

A related point has to do with that currently popular pejorative—greed. We've come a long way from a time when some people actually believed the famous line, in the original 1980s movie *Wall Street,* that "greed is good."

Greed is, in fact, a major offense against fairness. It involves an unconstrained egotism that is oblivious to the needs—and merits—of others. It means taking more than your fair share. And, as we have so often witnessed lately, unrestrained individual or corporate greed can be a major cause of "harm" to the rest of us. Greed is good mostly for the greedy.

So it is important to reserve the term merit for outcomes that are earned. A person's rewards should ultimately reflect contributions that benefit others, or the general welfare. The merit precept stakes a moral claim; it poses the right question. However, there is no simple formula for determining what is fair. Public norms, the marketplace, our representative democratic government, and our complex legal system can and do play a vital role in the imperfect art of determining what is merited in various contexts. (The important social science research domain known as "equity theory" also has much to say about how to achieve equitable outcomes in the myriad of everyday, real-world fairness decisions.)[19]

In any case, under a biosocial contract the merit precept is concerned with how to fairly distribute the economic surpluses of a society—the excess beyond what is required to provide for our basic needs. As I noted earlier, mutual aid has been a part of our species' formula for success for millions of years, and we both dishonor our past and defile our future as a species if we don't insist on it. The provisioning of our basic needs is our first responsibility, and differential rewards for merit must be confined to our "discretionary income" as a society—to use a textbook term—or else it converts the social contract into a zero-sum game.

Does this imply that we should turn the Reagan-inspired myth of "welfare queens" into a reality? Does it invite a culture of freeloading and an indolent class of economic "defectors," to use the game theory terminology? The answer is emphatically not. Where's the justice in that? In fact, a crucial corollary of our first two precepts is that the collective survival enterprise has always been based on mutualism and reciprocity, with altruism being limited (typically) to special circumstances under a distinct moral claim—what I refer to (once again) as "no-fault needs." So, to close the loop, a third principle must be added to the biosocial contract, one that puts it squarely at odds with the utopian socialists, and perhaps with some modern social democrats as well.

In any voluntary contractual arrangement, there is always reciprocity—obligations or costs as well as benefits—and all parties are net beneficiaries, even if the "goods" are not equally divided. As I noted earlier, reciprocity is a deeply embedded part of our social psychology and an indispensable mechanism for balancing our relationships with one another. Recall how everyone from Confucius to Cicero, Alvin Gouldner, and Ken Binmore, not to mention the horde of Golden Rule proponents over the centuries, have treated reciprocity as a sacred moral principle. Without reciprocity, the first two fairness precepts would look like nothing more than a one-way scheme for redistributing wealth—naked socialism.

When you think about it, the principle of reciprocity is also a fundamental aspect of any viable market economy. To repeat a point made earlier, this is what "exchange" and "trade" are all about. In a modern economy, our ubiquitous forms of symbolic exchange, from paper money to credit cards and electronic financial transactions, often obscure the underlying human relationships and distance us from one another psychologically. Nevertheless, these relationships are real, however attenuated they may be. Our markets may not be fair in terms of either equality or equity, but they do thrive on reciprocity.

A clear implication of this precept is that we are obliged to contribute a fair share to the collective survival enterprise in return for the benefits we receive. This directive applies to the rich and the poor alike—to both wealthy matrons and welfare mothers. We have a duty to reciprocate for the benefits our society provides. Otherwise we are in effect free riders on the efforts of others; we turn them into involuntary altruists. This is one reason why, for example, "workfare"—work requirements for public welfare recipients—was such a hot-button issue back in the 1990s and why the passage of welfare reform legislation mandating a work requirement for welfare beneficiaries defused this contentious debate, even though it caused some new problems (like providing adequate child-care services to the poor).

Equally important, however, is that being able to work is also a strong personal preference for the overwhelming majority of the poor—like everyone else. There is a large body of research in psychology and sociology documenting that unemployment can be profoundly depressing. Living on handouts is also a demeaning and humiliating experience for most of us, whereas doing meaningful work and being of use to others has intrinsic value and positive psychological effects, as the many millions of idled workers in our society these days are learning the hard way. Very few of us have a genuine preference for being unemployed.

Libertarians might object to such incursions on their freedom, but in my view John Rawls's definition of fairness under a social contract provides a definitive rebuttal: "The main idea is that when a number of persons engage in a mutually advantageous cooperative venture according to rules, and thus restrict their liberty in ways necessary to yield advantages for all, those who have submitted to these restrictions have a right to a similar acquiescence on the part of those who have benefited from their submission."[20]

However, the reciprocity precept also provokes a question: How can the differences in our ability to contribute be weighed and calculated? As is the case with merit, there are no easy formulas, but societies have worked out many ways of making collective decisions—from market transactions to legislative mandates, tax codes, charity work, and many others. As we shall see in the final chapter, developing creative new ways to contribute is currently a national priority.

Finally, I should note once again that the "dark side" of equity and reciprocity is also an important aspect of any biosocial contract. This concerns "letting the punishment fit the crime"—imposing sanctions for cheating or free riding. The alternative of turning the other cheek works only when there is genuine remorse and a passion for penance (the prospect of a voluntary behavioral change). Otherwise, policing the biosocial contract is at once a matter of necessity and a matter of fairness.

Conservatives and libertarians, many of whom are in denial about the idea of social obligations or universal social rights, might want to stigmatize this formulation as "fuzzy Marxism." But in fact the opposite is true. Marxism is grounded in a fuzzy biology, along with a simplistic and one-sided model of human nature. Marxism actually violates the fairness precepts outlined above. For one thing, Marx was diffident about specifying the content of our basic needs. To repeat, he allowed the inference to be made that equality and equity are equivalent. In a biosocial contract, this is emphatically not the case.

Marx also made no provision at all for merit and was quite hostile to capitalism. Remember how capitalists were viewed as the villains who were destined, in accordance with the imperatives of Marx's dialectical materialism, to end up in "the dustbin of history" (to use Bolshevik Leon Trotsky's epitaph for the Mensheviks). Most important, Marx's admonition that everyone should contribute to society in accordance with "ability," in the absence of the other two precepts described above and especially the claim for merit, could potentially be exploitative. Marx certainly did not frame his argument in terms of reciprocity, except during the transition to communism (see chapter 6, note 68). Despite some similarities in phrasing, the Marxian contribution principle does not accord with the biosocial contract paradigm.

Many other questions are begged by this ideological framework, of course, and some important qualifiers should be added. I will highlight and briefly discuss just a few of them.

First, there is the problem of the "naturalistic fallacy" (the false invocation of natural laws), dating back to Hume's famous essay. A critic might ask, Why ought we to care about our survival and reproduction, much less that of anyone else in our society? More to the point, why should anyone—especially the "haves"—accept a fairness ethic as a standard for guiding the policies and practices of a society? Even if we have been "programmed" by our evolutionary heritage to be concerned about fairness, how can anyone claim this creates a moral imperative?

Actually, these are the wrong questions. They amount to a sophist sand trap. The issue here is not whether we can justify some categorical imperative for morality. The reality is, we do care. Given the cardinal facts that we care intensely about satisfying our basic needs; that these needs must, by and large, be satisfied through the vast network of cooperative activities associated with the collective survival enterprise; and that we do, after all, have a shared sense of fairness, then the precepts of a biosocial contract provide a compass for steering a society through the political shoals. They provide a set of prudential normative principles that direct us to navigate a middle course between free market capitalism and egalitarian socialism. Moreover, these precepts represent existential imperatives in the sense that serious maladaptive consequences—both individually and collectively—will result from ignoring them and pursuing an alternative course (see chapter 5).

Second, how do we implement this ideology? How do we go about ensuring that our basic needs are met? Does this imply an economic and social revolution of some sort? The answer is yes and no. It implies a need to reform an evolved economic system that has many virtues but also some serious deficiencies. There are currently many effective market-based instrumentalities for meeting our basic needs. These must be augmented, improved, and supplemented; a better balance is required. (I will return to this point in more detail in the final chapter.)

Third, what about freedom, or liberty—a core value of Western democracies? The response to this concern is that we need to move beyond the naive assumptions and self-serving rhetoric about freedom (especially freedom for capitalists) that are so prevalent today. As the social critic Charles Morgan put it long ago, "Liberty is the room created by the surrounding walls." In other words, freedom always has boundaries. What we are talking about here is an adjustment in the location of the walls. For some, the room will be significantly expanded. For instance, more income for the poor would free them from some deep anxieties and severe, even life-threatening or life-shortening economic constraints. (Recall the lower average life expectancy for the poor noted in the previous chapter.) For others among us, there will be some shrinkage of freedom, but it would most likely be at the margins. In any case, this is a trade-off we must make for the long-term viability of the collective survival enterprise. It is in our common interest to do so, and it is only fair.

Fourth, how does a fairness ideology affect our supposedly sacred property rights? As I noted earlier, property has no unqualified natural right beyond what the rest of us are willing to recognize—and protect. This observation goes back to Plato and was seconded by Bentham, Rousseau, Locke, Marx, and many others, including the Supreme Court of the United States. In the real world there are many political constraints on property rights. They are reflected in such things as eminent domain, restrictive zoning and building codes, fire codes, condominium covenants, property taxes, curfews, and the like. Under a biosocial contract, property rights are further limited by what is compatible with meeting the basic needs of the rest of the population.

Fifth, what about those who cannot contribute their fair share of productive capital or labor to society? Is there some danger that unconstrained help for those who are truly needy would turn society into a vast charity ward, imposing an enormous economic burden on the rest of us? The fact is that we already willingly support our dependent children, the elderly, the disabled, and aging veterans, among others. We are more grudging as a society about aiding the poor and their children, because we know that many of them *can* contribute in various ways.

Nevertheless, there remains a hard core of people in our society who, for one reason or another, will always be unable to contribute: they are victims, not wastrels. Workfare, or "welfare-to-work" programs, will never work for them. This is a biological reality that we seem reluctant to face up to. But if our evolved moral sensibilities can encompass the victims of highly visible disasters like floods, earthquakes, terrorist attacks, and recessions, there is no moral ground for excluding the less visible tragedies all around us, including those that are, sadly, biologically based. (Down syndrome is one obvious case in point.) This is where the Golden Rule, and perhaps Rawls's "veil of ignorance," could be applied, especially since we ourselves, or someone we love, could also end up in need. Indeed, one of every five families in the United States has a member who suffers from some form of mental illness. Thus our social welfare programs also represent a form of social insurance for everyone, a public good that goes back to the Greek funeral societies.

Sixth, what if the existing economic and political order fails to provide for our basic needs? The historical record warns us that all regimes are ultimately contingent. Indeed, the American Declaration of Independence contains an enduring justification for breaking the "political bands" of the existing order. Governments "are instituted among men" to secure our "inalienable rights," and they derive their "just powers" from "the consent of the governed." Whenever any form of government "becomes destructive of those ends," the people have the right "to alter or abolish it." Plato and Aristotle warned that no political order is immutable. And the modern game theorists, whose research has done so much to illuminate the foundations of social cooperation in evolution, have shown unequivocally that mutual benefits are an essential requisite for a viable social contract. Defection is the likely response to an exploitative, asymmetrical interaction. To be sure, in human societies coercion is often used to prevent reforms, but the costs and risks are always high, and the long-term results are always problematic. A society may ultimately awake from its "social amnesia," as Barrington Moore called it. To reiterate, extremes of wealth and poverty have historically been the seedbeds of revolution.

Seventh, what about the objection that a fairness ethic can actually be unjust? This discordant note was sounded by two economists (one also a lawyer), Louis Kaplow and Steven Shavell, in their provocative 2002 book *Fairness versus Welfare*. "Contrary to the conventional wisdom," Kaplow and Shavell claimed, the notions of fairness and justice, especially as they are used in the judicial realm, can run counter to an individual's welfare. Accordingly, they asserted that social decisions should be based "exclusively" on welfare criteria and not on any legal principles. "The design of the legal system should depend solely on concern for human welfare."[21]

They do have a point. Our formal justice system can sometimes become so entangled in its legal principles and precedents that the substantive issues get obscured or short-changed. As Charles Dickens's frustrated Mr. Bumble put it, sometimes "the law is a ass—a idiot." There is also the line spoken by an apprentice lawyer in a movie about the life of Jane Austen, "Justice plays no part in the law." This is most dramatically the case with the doctrine of strict constructionism, which requires a judge to apply the text of the United States Constitution only as it is written (once a court is clear about the meaning of the text, no further investigation of the merits is required), as well as the doctrine of "originalism," the view that the Constitution should be interpreted according to the (supposed) intent of those who wrote and adopted it.

Slavish adherence to such judicial principles can produce a "trained incapacity" to consider the merits of a case (in economist Thorstein Veblen's ironic phrase). We have recently seen an appalling example of this in the Supreme Court's decision to apply the "freedom of speech" clause of our Constitution expansively to include corporate spending on political campaigns, which may well serve to drown out the free speech of the rest of us. Justice John Paul Stevens, in his ninety-page dissent, suggested that the majority of the court had lost touch with the notion of fair play.

However, the Kaplow and Shavell thesis ultimately founders because they insist on defining fairness in a narrow, legalistic way that contradicts common usage and common sense. They confine the term to judicial principles and practices that *ignore* and even conflict with an individual's wellbeing. "We argue that legal policy analysis should be guided by reference to *some* coherent way of aggregating individuals' well-being, in contrast to the view that [it] should be guided by notions of fairness."[22] As examples of such legalistic myopia, they cite such things as "corrective justice" (or compensating someone you have harmed), "promise keeping" (adhering to contracts), and "retributive justice" (proportionate punishments for those who cause harm).

As "fairness" is used here—and as most nonlawyers, I believe, commonly use the term—it is concerned precisely with well-being. However, fairness is not just about the well-being of individuals. This is a key difference from Kaplow and Shavell's usage. Fairness is all about adjudicating among the many, often competing, interests and welfare claims, as I noted in chapter 2. Moreover, fairness is always concerned with the substantive consequences, not only for the parties directly involved but for the well-being of society as a whole. The three fairness precepts included in the biosocial contract are, I believe, decision rules that transcend individual welfare and speak to the larger concern for our general welfare.

President Obama was criticized for using the word "empathy" to characterize the temperament he was looking for in a Super Court justice during the nomination process for Sonia Sotomayor. A better term would have been "a deep sense of fairness"—empathy for everyone.

Finally, there is the politically explosive issue of where to draw the line, or lines. Is it realistic to have an open-ended commitment—an entitlement—to provide for the basic needs of all potential claimants? Should we accommodate an unrestricted number of babies born to welfare mothers, or deadbeat fathers? Would this include the continuation in perpetuity of an open-door immigration policy, or an unending flood of illegal immigrants? And how do we draw lines in a global economy, where more and more of our needs and wants are satisfied by underpaid workers in other countries? Global poverty is a vast ocean of unmet needs. For example, in Mexico alone, some 40 percent of the population of about 100 million live in deep poverty. There are no easy answers to these questions, but I would reiterate a key point made earlier about the nature of the collective survival enterprise. It is based on mutualism and reciprocity, not altruism. So the general answer to the question is that, to be consistent with the precepts of a biosocial contract, lines will have to be drawn. This is an inescapable trade-off. (I will return to this issue in the final chapter.)

It is worth repeating the simile used in an earlier chapter. Fairness is like a "golden thread" that binds together a harmonious society. And when this thread breaks, the social fabric unravels. The response to the former British prime minister Margaret Thatcher's contemptuous claim that there is no such thing as a society (only individuals and families) is that, on the contrary, a society exists when people believe it does and act accordingly—and vice versa. But fairness is not an all-purpose formula or recipe. It's a general principle that recognizes the claims of competing interests and directs us to find fair compromises—equality, equity, and reciprocity—as appropriate.

In this paradigm, compromise is not, as a rule, a sellout of one's principles to political expediency. Compromise is a principle with a higher moral claim when fairness is at issue. The spirit of compromise recognizes and accommodates legitimate competing claims and interests, and it serves the overarching goal of preserving a cooperative society, as Garrett Hardin perceptively argued. However, the evidence is all around us that fairness is often a matter of perspective: it can involve a very difficult judgment call. That is why we have a formal justice system, and mediators, family counselors, contract negotiations, legislatures, and—not least—markets. Indeed, every stable society has a tool kit of customs and practices for approximating fairness, as I noted in chapter 2.

However, social justice can be specified, to a first approximation, within the framework of a biosocial contract. It is grounded in the bedrock imperatives of our basic needs, using the objective measures provided by the Survival Indicators program. A fairness ideology provides both a biological justification and, ultimately, a political imperative for striking a better balance between providing for our basic needs and rewarding merit. More important, it provides specific (measurable) criteria for where this balance can be found.

We conclude, then, by returning to where we began (as a species). Charles Darwin recognized that a human society is quintessentially an interdependent survival enterprise. And so did Plato and Adam Smith and Edmund Burke. The superorganism is vital to our survival and reproduction. However, this vision of our collective purpose does not negate or ignore our individual self-interests. Rather, it represents an aggregation of those interests into an immensely complex system of synergies based primarily on mutualism and reciprocity. The poet John Donne's immortal line "no man is an island" is true in a very practical, economic sense.

Accordingly, we can also reaffirm that the modern democratic state has evolved as an instrumentality for self-government and the pursuit of our common needs (the public interest), although its purpose has all too often been subverted. Plato and Aristotle apprehended this basic purpose in their conception of the *polis,* and they prescribed a mixed government under law as our best hope for ensuring that the public interest would be served. Plato and Aristotle also recognized that a fair-minded form of justice is an essential element of the public interest. This is the only way to ensure longer-term political stability and social harmony.

What the biosocial contract adds to Plato's great vision is the recognition that there are in fact three distinct categories, or types, of substantive fairness and that these must be combined and balanced in appropriate ways. The substantive content of social justice consists of providing for the basic needs of the population, along with equitably rewarding merit and insisting on reciprocity. Indeed, appropriate rewards for merit provide important incentives for the vast combination of labor that is required to meet our basic needs. The biosocial contract also enlists the growing power of modern evolutionary biology and the human sciences to shed light on the matter, and it identifies an explicit set of criteria for reconciling (if not harmonizing) the competing claims that have been promoted by political ideologues of the left and the right.

I believe this framework offers our best hope for achieving and maintaining that elusive state of voluntary consent that is the key to a harmonious society—a Nash equilibrium writ large. This is an ideal worth striving for, because our own survival, and more certainly that of our descendants, may well depend on it. We should adhere to the "seventh generation" law of the Iroquois: Everything we do should be viewed in light of its impact on future generations. As the great American public park designer Frederick Law Olmsted put it, "The rights of posterity take precedence over the desires of the present." Nothing less than our evolutionary future is at stake.

8 * The Future of Fairness: The Fair Society

> Seeing every man, not only by Right, but also by necessity of Nature, is supposed to endeavor all he can to obtain all that is necessary for his conservation, he that shall oppose himself against it, for things superfluous, is guilty of the war that thereupon is to follow.
>
> THOMAS HOBBES

> The rich must lead a simpler life so that the poor can simply live.
>
> MAHATMA GANDHI

Nobody can say we haven't been warned. From Plato and Aristotle to the latest game theory models, it should be clear by now that a harmonious society depends, absolutely, on fairness and social justice. This is surely the "old truth" that, as Goethe claimed, the "wisest spirits" have long understood. (Some other wise spirits are quoted in the endnote.)[1]

However, as a nation we seem to have forgotten this old truth. In 2008 the Organisation for Economic Co-operation and Development, a consortium of thirty advanced democratic nations, reported that the United States had both the highest level of income inequality and the highest poverty rate among all the OECD members except Mexico and Turkey.[2] During 2009, you may recall, almost 50 million of our people, including 17 million children, also experienced "food insecurity" (hunger) at some point. Now the situation is even worse, with massive job losses, declines in average working hours, home foreclosures everywhere, and an ever-widening sea of desperate poverty.

An alarming wake-up call by Bruce Judson of the Yale School of Management in his 2009 book *It Could Happen Here: America on the Brink* is only the latest of many recent warnings that our society has entered very dangerous waters politically.[3] (We have glimpsed a looming iceberg ahead in the recent vitriolic health reform debate and what promises to be an ugly off-year election.) Wall Street may be in recovery, but Main Street is still in a full-blown depression with no end in sight (as this book goes to press). How can anyone justify this?

It is abundantly clear that we need a major change of direction. We need to set a new course toward a society that adheres to the precepts of a biosocial contract. Free market, laissez-faire capitalism has been fully "field tested," and it has failed in its promise to provide the greatest good for the greatest number. Without a tight leash and the leavening of other social values, capitalism systematically favors the rich and is cavalier about our prime directive. Government, moreover, can become a maidservant rather than a counterbalance.

Yet, as we have seen, the socialist model is not the answer. Nor, certainly, is the libertarian fantasy. Utopian schemes of all kinds are based on faulty assumptions and wishful thinking about human nature, along with a tacit denial of our inescapable survival challenges. Nor can we rely on an anemic Third Way that has proved unable to confront the status quo.

What the science of human nature teaches us, I believe, is that the only sustainable alternative going forward is what could be called the Fair Society model. We cannot build a stable society based solely on an unconstrained pursuit of self-interest, or on an egalitarian altruism, or even (unaided) on our easily subverted sense of fairness. We must, instead, work toward achieving a self-enforcing (and enforceable) biosocial contract that strikes the proper balance between the three fairness precepts I discussed earlier—equality in relation to our inescapable basic needs, full and fair recognition for merit, and proportionate reciprocity. (To borrow a metaphor from sociologist Amitai Etzioni, we need a three-legged stool.)

This, as the British like to say, is the "Big Idea." If we were to fully implement the concept of a Fair Society, I believe it would transform the future of this nation.

The first (and most urgently needed) leg of the stool is an unqualified commitment to our prime directive. This requires what I refer to as a basic needs guarantee (or "bio-socialism"). This is not socialism with a hyphen, or a face-lift. It entails a major ideological shift away from traditional socialist and social democratic values to something that is more focused on the biological imperatives documented in chapter 5, as well as being more realistic and fair-minded. Fairness is not about creating an egalitarian society that ignores our many differences. It is not about an equal sharing of the wealth, much less dissolving ourselves into small autonomous communities. To repeat, it is about taking collective responsibility for the basic biological needs of all members of our society. This is our heritage as a species and, I submit, it is also our deepest moral obligation to future generations.

However, a basic needs guarantee (biosocialism) is not simply concerned with mending our safety net so that we are better able to cope with calamities. This is only part of what must be a much broader agenda. The goal is to ensure that the full range of our common biological needs is satisfied in a continuing way, especially the needs of the next generation. We must reverse the steeply negative trends in our economic, health, and other vital statistics, as highlighted in chapter 6.

This implies a commitment to positive social development and a cradle-to-the-grave "package" of goods and services (from adequate prenatal care to end-of-life care). It includes such basic resources as ample (clean) freshwater, clean (renewable) energy, low-cost transportation, efficient waste removal (and recycling), affordable health care (plus, let's face up to it, mental health services), safe streets, adequate child care, and free lifetime public education (because it ultimately benefits our society in many ways), along with such traditional basics as wholesome food, adequate shelter, and a healthful environment.

A cradle-to-the-grave basic needs program would not have to start from scratch, of course. It would build on our existing private enterprise economic system and a century of public sector social welfare experience in this country, not to mention our vital private and religious institutions and charities. However, it would also require a holistic approach that encompasses all fourteen of our basic survival and reproductive needs, and it would focus on achieving the kind of measurable outcomes discussed in chapter 5.

The single most important first step toward achieving this end must be to provide employment to everyone who is willing and able to work. This is an older social ideal that has been largely abandoned as a policy goal in this country during the past thirty years, but it has gained new urgency during these difficult times. Currently there is only one job available for every six job seekers, and the projections for private-sector job creation for the next several years are dismal. The Labor Department expects that eight out of every ten new jobs will be "low-paying" or worse. (Here again, some European countries have done a much better job of coping with the economic crisis.)[4]

Employment for everyone who is able to work—at a living wage and not our delusional minimum wage—is by far the fairest and most dignified way to provide for many of the basic needs of our people. And where the private sector cannot generate good employment, a more robust public-sector employment program should provide a backstop as the "employer of last resort." This concept traces back to the original version of what became the Employment Act of 1946, although it fell into disfavor among economists in recent decades. As I noted earlier, thanks to the work of Milton Friedman and others, many economists came to believe that full employment would cause inflation. (See endnote 5.)[5]

I am not talking here about creating "junk jobs" or "survival jobs"--they must be socially useful jobs. This could provide many additional benefits for our hard-pressed states and local communities, including much-needed public-service jobs and a reduction in safety net financial burdens for the states. At the same time, it would satisfy the criterion of reciprocity. The benefits would be earned.

Of course, it's also important for the government's role to be more countercyclical in nature and not compete with the private sector. Accordingly, there should be even greater incentives than already exist for private-sector job creation and more emphasis on training workers for the twenty-first-century job market. The Obama administration has in fact initiated some job-creation measures beyond the original economic stimulus package, but this is only a start.

A corollary of a full employment policy is that we also must devise a way of paying workers a living wage. One intriguing idea was proposed by senior economist and Nobel laureate Edmund Phelps in his 1997 book *Rewarding Work.* Phelps argued that we should increase wages in a way that does not impose an unrealistic and destructive cost burden on our nation's employers. The idea would be to supplement the minimum wage, and low wages generally, with something like an automatic rebate of payroll taxes that employers would then pass through to their employees. Not only would this reinforce the value of work and greatly improve the lives of the one-third of our population who are trapped in deep poverty, but it would move us away from a welfare system that is demeaning and inefficient and that does not focus effectively on rewarding work. Among the obvious objections to this plan, apart from its cost, is the danger that employers might try to game the system to their own advantage. But Phelps deals with this and other issues at length in his book.

Why not simply give everyone enough money to meet their basic needs? In fact, this is precisely what has been advocated for many years by the proponents of the basic income guarantee (or BIG)—a generic name for various proposals in various countries that are designed to provide a universal income floor for everyone, with no strings attached (as a "social right"). In other words, there would be no work requirements—though some versions of the BIG call for a vaguely defined social contribution. Supposedly a universal basic income would eliminate the need for a welfare system with "caseworkers," as well as the many separate income replacement or supplement programs for the unemployed, the disabled, the working poor, and others. BIG promoters claim such a program would be simple to administer and, at a single stroke, would eliminate absolute poverty.[6]

The idea is not so far-fetched or idealistic as it may seem. Our categorical cash benefit programs already accomplish a significant part of this objective, although there are many holes in the system (for example, when the often inadequate unemployment benefits run out). Back in the 1960s there was a serious effort to enact a guaranteed income program in this country, which had the backing of over twelve hundred prominent economists and many liberal politicians. But the idea died with the Vietnam War and the conservative political revolution in the 1980s. Today it lives on mostly among socialist and "green" parties in Europe and the United States, as well as in a few other liberal enclaves in this country.

Apart from the fact that such handouts would be a very hard sell in our more conservative political era, it also violates the reciprocity precept and would create a major "policing" problem to prevent free riders and cheaters—those who are not in need—from exploiting the rest of us. For all their faults and inefficiencies, our categorical income replacement or supplement programs have the virtue of being targeted to demonstrable no-fault needs and are conditional about the benefits provided. This is more consistent with the precepts of a biosocial contract.

Our income-oriented social welfare programs could obviously be greatly improved. For instance, the benefit levels are in general very inadequate, especially in comparison with those of other industrialized countries, and the eligibility rules are often restrictive. There are also many millions of unemployed people who are currently without income from any source. (It is estimated that about one-third of those why are eligible for unemployment benefits never receive them.) They are falling through the cracks. Also, we must fully acknowledge that there will always be many people who are unable to contribute a fair share—our young children, the extremely old, those who are too sick or severely disabled, the developmentally impaired, the involuntarily unemployed, and the damaged souls of our society.

Yet even a full employment program at a living wage would not be sufficient to provide for the full range of our instrumental needs in an age when a loaf of bread can cost the better part an hour's work at the current minimum wage. A major corollary of a full employment policy and improved income replacement/supplement programs, therefore, must be a far more concerted effort to eliminate financial barriers that prevent access to such essential goods and services as housing, health care, higher education, child care, and transportation. Here we have much to learn from other welfare capitalist societies, where programs such as family allowances and educational benefits help parents support the extra cost of having children, and where there is good-quality, low-cost housing, excellent mass transit systems, free health care, and even free education. (The much-touted earned income tax credit in our society amounts to much less than meets the eye. See the endnote.)[7]

Consider this: According to the College Board, some 65 percent of all the four-year college students in this country who graduated during 2006–7 were burdened by an average of $23,800 in college loans (for private college students the average was estimated to be much higher, at $38,300), and this didn't even include credit card debt. Overall student indebtedness was probably close to $35,000 on average, and now the problem is even worse. In other words, the best and the brightest of the next generation of Americans are being saddled with a heavy load of debt at a time when their earnings are relatively low (if they can get a job at all), when they are just starting families, and when they should be saving to buy a house. This is a form of national insanity. Surely we can do better than this as a society. (For more, see the endnote.)[8]

If we had a social welfare system that was more comprehensive, more adequately funded, and more readily available to all our citizens (less restrictive), it would truly be transformative—provided, of course, that we also make major quality improvements. Much innovation, new ideas, and appropriate incentives—both carrots and sticks—would also be needed.

To cite just one example, the traditional approach to providing public housing was to demolish whole neighborhoods of older housing and build new apartment ghettos, all at great expense. Today we have many unused facilities—closed military installations, shuttered factories, and abandoned homes in profusion in some localities—that could provide low-cost housing and homeless shelters. But there must also be a quid pro quo for the recipients, namely, responsible behavior and a willingness to work and contribute wherever possible. (A case in point are the privately supported housing programs, like Habitat for Humanity, that expect their beneficiaries to contribute labor.) Otherwise the enhanced benefits would amount to uncompensated charity. (More on this issue below.)

In an era when there is deep suspicion, even hostility, toward government in some quarters (witness the tumultuous, angry, even violent public debate over health care reform), how can we achieve the broad changes that a basic needs guarantee (or biosocialism) would require? Who should be responsible for producing the needed changes? Some of us will strongly object to a larger government role. After all, private charity is already robust and has a widespread influence in our society. Why not rely on our established religious and civic organizations and private foundations? Or perhaps leave it to good-hearted individual benefactors to decide how best to use their money? Bureaucracies of any kind can siphon off much of the resources intended to help those in need.

The answer is that there is no one best way to get the job done. The challenges out there are so many, and so urgent, that we need to sound General Quarters and call all hands to their battle stations. Private charities and individual donors will continue to play an important role going forward. However, there must be a shift of priorities in private giving toward alleviating poverty and breaking the poverty cycle. Also, realistically, the scope of our unmet basic needs as a society goes far beyond what private, charitable resources can satisfy.

In fact, there are many essential public services that can be provided (effectively) only through collective action—from garbage disposal to police and fire protection, electric power, and of course health care. Sometimes the private sector can accomplish the job. Cooperatives and nonprofit organizations are especially popular with many of our public utilities, like local power companies and solid waste removal. On the other hand, private-sector service providers can sometimes be aggressively profit-focused, to the point of deliberately excluding those with low incomes or with costly needs. This is precisely why our predominately private health insurance system in this country has been at once highly profitable and seriously deficient in insuring people with low incomes, or preexisting conditions, or major illnesses.

In such cases a government role may be needed to achieve a better balance between profits and service and to subsidize those who cannot afford to pay. Government can sometimes do a better job of spreading the risks and costs, imposing fees or taxes so that everyone pays a fair share, and ensuring that everyone receives the needed benefits. A fair-minded reading of the new health reform law provides a stunning example, whatever its shortcomings. Sometimes, in fact, government can do the job more efficiently and at lower cost. (Again, recall that the overall administrative cost of Medicare runs about 3–5 percent, compared with 25–30 percent, including profits, advertising, and lobbying costs, for private health insurers.)[9]

An especially inspiring example of what government at its best can do is Oportunidades, the Mexican government's boot-strapping program that pays that country's poorest mothers to provide good nutrition and health care for their children and, equally important, keep them in school. It's called the "Conditional Cash Transfer Program," and it has been hugely successful, with low overhead, low corruption, and stunning results. Some 25 million children are now participating, and a rising new generation of Mexican children is transforming a once deeply depressed underclass.[10]

The second major element of the Fair Society model—the second leg of the stool—involves reforming the private sector so it is tethered more closely to the underlying purpose of the collective survival enterprise. Earlier I argued that we should not destroy the dynamism and the social benefits that capitalism, at its best, can produce. Capitalism is preeminently a system that rewards various forms of "merit"—the second of our three fairness precepts. This remains a crucial moral value.

However, our society has the right to regulare the private sector. Markets are, after all, embedded in societies, and the conduct of the private sector should always be weighed in relation to its impact on the larger public interest; the profits that private businesses achieve are, in effect, rewards for merit—for the contributions they make to our society. (Of course, many capitalists and libertarians reject this formulation, but this principle is well established. We have the power, if we choose to use it, to insist on it.)

Moreover, profits are not all created equal. Cigarette company profits are an obvious negative example, and so are the outrageous bonuses received by investment bankers. It is now abundantly obvious that free market capitalism in this country serves very well the "greatest good" of the few at the top of the pyramid and has short-changed the greatest good of a much larger number. So the question is, How can we realign capitalism—and the merit precept—more closely with the public interest?

It has been customary in our society to assume that those who put up the capital to finance our business ventures are also entitled to be the owners and controllers or managers, and there are many laws, precedents, regulations, and customs (not to mention economic theories) that support this nexus. But in fact there is no necessary connection between capital, ownership, and control. These three functions can be decoupled, and there are a great many real-world examples to prove it. For instance, many public corporations have nonvoting forms of stocks and bonds, and much of the short-term capital financing that is provided to our businesses, large and small, comes from bank loans, where the obligations are limited to repaying the principal with interest.

Mainstream economic theory during the past three decades has stressed the primary importance of "private goods" and the objective of increasing "shareholder value" and maximizing profits for the company's "owners," even if they are only nominal owners. But there is a very different way of looking at the modern business organization. I like to call it the "corporate goods" model. As I mentioned earlier, corporate goods refer to any economic benefits that are cooperatively produced by two or more autonomous "players" working together—investors, managers, line workers, joint-venture partners, suppliers, distributors, retail dealerships, even the government—where the benefits can be divided up in any number of ways. Who benefits from these corporate goods and how the profit pie is "distributed" is ultimately a matter of social (and political) choice, as Galbraith points out.

As a rule in our capitalist system, the owners or managers and (maybe) boards of directors decide how the company should be operated and how the pie should be divided. But what if the "payoffs" in the game were subject to negotiation among the players? What if all the stakeholders were empowered? What if the Platonic principle of mixed government was applied to the governance of private companies, with the CEOs answerable to all the stakeholders?

In this model, the participants would all have a right to share in the control and management of the organization (in appropriate and not disruptive ways, obviously) and to benefit proportionately—equitably. Consider this: What if our CEOs were compelled to consult with the other stakeholders in the company (like the workers) regarding their compensation rather than entrusting it to a lapdog board of directors and professional "compensation consultants" with a built-in conflict of interest? It's a heresy, of course, but an intriguing idea.

There are, in fact, many examples in our society of enlightened companies that have moved in the direction of a stakeholder model. Corporations like Microsoft, Google, and Apple provide generous perks to their employees. Many more companies have various forms of employee stock ownership and profit sharing arrangements, not to mention 401(k) retirement plans. Most of this largesse represents textbook examples of "modern" personnel practices. Independent studies of what has been labeled "shared capitalism" indicate that such measures are beneficial both to the workers and to their employers.

Worker suggestion boxes have been a fixture in large organizations for generations, and management theory as taught in our business schools these days stresses such themes as "self-managed teams," "decentralized decision making," "worker empowerment," "consensus building," "flattened organization," and even "servant leadership." But all of this relates primarily to internal management techniques and practices.

What has been termed "stakeholder capitalism" involves something more. It implies a greater role for various stakeholders in the control—the overall planning and governance—of a company and (most important) in the basic values that govern its actions. Stakeholder capitalism is preeminently a way to change the power balance in capitalism, in the interest of achieving fairness toward all the stakeholders, including society.

Though the term itself is only now gaining in popularity, the concept of stakeholder capitalism is not new. Its roots can be traced back at least to the "corporatist" industrial model first developed in Germany in the 1870s, where workers are given seats on corporate boards under a policy called "co-determination" and where the government may also exert an influence on overall corporate policy.

A somewhat different model was developed by the Japanese after the Meiji Revolution, which led to Japan's rapid industrialization during the nineteenth century. The *zaibatsu* (now called *keiretsu*) that still dominate some large Japanese companies today are family-owned firms that have reciprocal ownership arrangements with other institutions, like banks and insurance companies, and are also subject to strong government "guidance."

But perhaps the best historical model for stakeholder capitalism can be found in the United States itself. Though mostly forgotten now, in the decade or so immediately after World War II, when there was still a carry-over of wartime patriotism and a strong sense of community in this country, large American corporations commonly practiced a "good neighbor" policy, and their CEOs were viewed as "corporate statesmen." There were no formal institutional distinctions, but managers and boards of directors were highly attuned to maintaining good public relations and keeping their workers happy, and they generally acted accordingly. Moreover, the public expected socially responsible behavior and punished firms that were tone-deaf. In other words, a different kind of corporate and societal culture existed, and this significantly influenced the behavior of our large business firms. (Robert Reich calls it a "not quite golden age" in *Supercapitalism,* because it was certainly not a golden time for blacks and women.)

One indicator of the close relation between public expectations and the way private industry conducts itself is the findings of an international public opinion survey back in the mid-1990s regarding public perceptions of corporate ownership. When survey respondents were asked "Whose company is it?" (stakeholders' or shareholders'), the percentage breakdown in Japan was 97–3 in favor of the stakeholder model. In Germany the percentages were 83–17. However, in the United States and the United Kingdom the results were reversed. It was 24–76 percent and 29–71 percent, respectively, in favor of the shareholders. These differences in public attitudes correspond to cross-national differences in corporate cultures and in corporate behavior.[11]

In more recent decades, the idea of stakeholder capitalism has had its ups and (mostly) downs in this country. The general idea, if not the term itself, was widely touted during the 1970s and 1980s by management gurus like Peter Drucker, W. Edwards Deming, Tom Peters, and Russell Ackoff. Their basic argument, in a nutshell, was that private businesses are human organizations that, in turn, are parts of the larger society. They do not exist in a moral, economic, or political vacuum. Whether or not they recognize it—and they certainly should—large businesses all have large networks of relationships and constituencies.

Some proponents of stakeholder capitalism have tended to treat the idea as a stark alternative to shareholder interests. But this adversarial view is a false dichotomy. Shareholders obviously are also vital, especially in an era of large capital needs and fast-moving global financial markets. Ideally the shareholders and the other corporate stakeholders should have a shared sense of purpose and a close alignment of interests. In any case, stakeholder capitalism would address the fundamental moral problem with conventional capitalism and provide a more equitable sharing of the benefits. (The pros and cons of the concept were fully debated in a 1997 book called *Stakeholder Capitalism.*)[12]

There are some outstanding examples these days of companies that try hard to be good neighbors, if not explicitly following the stakeholder capitalist model, such as Proctor and Gamble, Microsoft, Unilever, Nordstrom, and others. But this is not the rule, and stakeholder capitalism is certainly not taught in our business schools, although there are signs that the climate may be changing. At Harvard Business School, one of the leading institutions for training our future business managers, about half of the 2009 graduates (over four hundred) took a public oath at an unofficial pregraduation ceremony at which they pledged to uphold high ethical standards. Among other things, according to the *Economist,* the graduating students promised to "serve the greater good" and avoid "decisions and behavior that advance my own narrow ambitions but harm the enterprise and the societies it serves."[13] A fine thing, if these future captains of industry actually practice what they've pledged. But we must also guard against the other half of the graduating class who didn't sign up and are, apparently, unrepentant capitalists.

In theory, stakeholder capitalism sounds straightforward. All the stakeholders in a company—investors, managers, workers, suppliers, joint-venture partners, customers, and of course the community at large—should have a voice in influencing key decisions, policies, and practices. The problem, needless to say, is how to implement this ideal in a way that avoids creating a managerial nightmare—a Gordian knot that makes the company rigid and inflexible, with various stakeholders focusing on their own interests and becoming obstacles to needed change. (Labor unions in this country, for instance, have too often played an adversarial role, although many companies have also provoked it.) The fear is that "democratizing" business firms will degrade their efficiency and effectiveness in our intensely competitive global environment. Globalization is the new reality that American companies back in the 1950s did not (yet) have to deal with.

There are certainly no all-encompassing answers to this dilemma, but an overall solution begins with a shared understanding among the stakeholders that their own interests and the future success of the company are closely linked. This implies a commitment by the stakeholders to share the pains as well as the gains. In practice, both capital and labor may need to moderate their expectations to ensure long-run viability for the company. A clear test case of this stakeholder model is the creation of the "new"

Chrysler and "new" General Motors. Having passed through the valley of corporate death, these companies have been compelled to make deals that give the workers and the government major ownership stakes, along with the bondholders in the case of GM, in return for various stakeholder sacrifices. A great many people will be keenly interested in the outcome of this crisis-driven experiment.

However, there are also some examples of a more voluntary path to stakeholder capitalism. One of the most outstanding of these, I believe, is the Organic Valley Family of Farms (and its CROPP farmer cooperative), a remarkable success story that may be the best thing that has happened to family farms in this country in the past hundred years.[14] Most farmers today, especially small farmers, are in thrall to the huge, powerful food conglomerates—General Foods, General Mills, Nabisco, Swift, Smithfield, and others—and are completely at their mercy when it comes to the prices they receive for their produce and the level of debt they are required to take on in order to be "efficient" and "competitive." Organic Valley is changing this exploitative dynamic.

In essence, Organic Valley is a kind of hybrid between a member cooperative and a textbook for-profit business venture. It was founded back in 1988 by a small group of Wisconsin farmers as a dairy cooperative, and it continues to this day to be controlled and managed by its farmer-members. Now numbering more than sixteen hundred farms nationwide with annual revenues of more than $520 million, Organic Valley is managed in a remarkably democratic fashion. The CEO is one of the founding farmers. The board of directors is elected by farmer-members, and each of the company's product lines (referred to as product "pools") is actively guided by farmer committees that are also elected. The CEO, George Siemon, likes to call it "a social experiment disguised as a business."

In many other respects, the administrative structure, with a staff of over five hundred, looks like any other for-profit organization. Indeed, it has an aggressive development strategy. But there are some important distinctions. One is that, instead of creating their own processing, packaging, and transportation facilities, Organic Valley has worked out an extensive network of joint-venture (contract) relationships with existing operators around the country. Another difference has to do with how the profits are divided up. First the company pays dividends to nonvoting class E stockholders who helped to finance the development of the company, along with paying other deferred debt obligations. After that, 90 percent of the remainder is distributed equally among the farmer-members and the staff, while 10 percent is donated to local community development projects nominated by farmers and staff members.

By far the most important achievement of Organic Valley, though, is the prices paid to the farmers. They are much more stable and much more generous than conventional agricultural prices. Take dairy prices. In 1989, conventional milk prices fluctuated around $12.30 per hundred pounds (abbreviated as cwt), while the one-year-old Organic Valley paid its farmers a flat, stable $14.30 per cwt. For the next eighteen years, the milk prices that conventional farmers received followed a sawtooth pattern that averaged about $13 but fell as low as $10.57 per cwt at one point, while Organic Valley prices followed a smoothly ascending curve (part of it to offset higher costs for organic farmers). In 2009, when conventional milk prices ended the year at $12.90 per cwt, Organic Valley farmers were receiving an average of $23.75. Even allowing for the higher expenses for organic growers, the difference is analogous to a living wage versus the minimum wage, given the slim profit margins for most small farms these days. In other words, the higher prices you may pay for Organic Valley products directly benefit our hard-pressed small farmers.

As is true everywhere else in the organic food business these days, sales and prices at Organic Valley are under tremendous pressure owing to the recession. The company has had to trim its sails a bit, but it is still on course with a stakeholder business model that promises a much brighter future for the beleaguered family farm in this country. It's a model that is both ecologically sustainable and economically fair for the farmers. CEO Siemon speaks of having a more complex business model with "a triple bottom line"—profits, planet, and people.

There are, to be sure, many issues and many subtexts related to the concept of stakeholder capitalism. Here I will mention just a few.

One of the most challenging problems, already noted above, is how to engineer greater stakeholder participation in management without diminishing the effectiveness of the organization. Organic Valley shows that there are responsible ways to do this, and that it can even improve the effectiveness of the organization. Indeed, many of our better-managed companies already have ways of networking with various stakeholders. Expanding and strengthening this model going forward is likely to be an evolutionary process, with a learning curve over time for developing new attitudes and new roles.

A second problem has to do with augmenting the voice of consumers and the public in corporate governance, beyond traditional market research and sales "signals." Consumers often need an ombudsman. Likewise, shareholders have only begun to take back control over the many rogue CEOs who have been lining their own pockets in recent years at the expense of their fiduciary responsibilities. In fact, the Securities and Exchange Commission recently announced a plan to give major shareholders greater ability to nominate candidates for corporate boards, a step in the right direction, along with the legislation that would allow for (nonbinding) shareholder votes on executive compensation packages. As a guest writer for the *New York Times Magazine* commented, "American companies lack not only capable managers but also patient and farsighted investors. . . . Neither managers nor shareholders have wrapped themselves in glory. Is it too much to dream that one day they might be forced to listen to each other?"

Another issue is sharing pain when times are bad, an especially acute concern during these troubling times. For instance, the all too common practice these days of locking out "surplus" employees without any notice or severance pay is fundamentally unfair. Another difficult problem, especially for the many millions of small businesses that operate on tight profit margins, is how to juggle the objective of paying fair compensation—a living wage—with keeping operating costs in line. A flattened compensation structure is an obvious starting point, and so is a more generous minimum wage that all businesses must pay. But there are no all-purpose answers to the problem without something comprehensive like Edmund Phelps's low-wage supplement.

Finally, how can a business be fair-minded toward its stakeholders and still survive in the face of global competition from countries with rock-bottom wages? This has been a formidable problem for American businesses during the past thirty years. Yet despite these challenges, the stakeholder capitalism model could certainly be greatly expanded if we make it our goal as a society, and as a global economy, to do so.

Needless to say, even a fully developed pattern of stakeholder capitalism, with a more equitable sharing of the profits, would still be insufficient to provide for the full range of our basic needs. Our deepening poverty and rising unemployment statistics attest to the continuing failure of capitalism as an all-purpose solution to our ongoing survival challenge. Another part of the solution may lie in the often underrated not-for-profit and cooperative sector of our economy, including what has been called "social entrepreneurship" and "community governance." Perhaps this could be viewed as adding another leg to the stool.[15] (I'll talk about the third leg below.)

Ever since cooperative founder Robert Owen's day, there have been a great many successful cooperative ventures in this country and elsewhere that are member-owned (or customer-owned)—grocery stores, credit unions, local power companies, grain elevators, insurance companies, and the like. They are referred to as 501(c) organizations in the United States because of the tax law that authorizes these and other nonprofits like our charities, where the revenues may come from grants, donations, member contributions, or even sales, and where any "surpluses" generated must be retained by the organization or returned to the members. In Spain, the sprawling cooperative Mondragón Corporation—the largest in the world—is also the seventh largest industrial enterprise in the country. These nonprofit organizations are sometimes more effective than capitalist businesses in meeting our basic needs these days, because their values are not oriented primarily to profits. And it happens that there are some exciting new developments—green shoots with only their cotyledon leaves showing—that hold great promise for the future.

I like to call it the "wonderful life model," from the classic Hollywood movie that immortalized the old-fashioned, member-supported building and loan associations. These community organizations provided financing for many low-income home buyers from the 1930s through the 1960s, and even today—though their role has been much reduced and much changed. The building and loan slogan for many years was "neighbors helping neighbors," and it was really a form of economic socialism. Mortgage or loan payments from members provided the wherewithal to finance the mortgages and other loans for new members. No wonder that, in film director Frank Capra's timeless morality tale, the nasty, mean-spirited, even dishonest capitalist bank president, Henry F. Potter (Lionel Barrymore), was hostile to the building and loan do-gooder George Bailey (Jimmy Stewart).[16]

Something similar in spirit to the classic building and loan model, under the heading "microfinancing," is beginning to happen in the small-business sector for low-income entrepreneurs without capital who want to open hair salons, gift shops, car repair services, and the like. The idea was pioneered in Bangladesh by the now world-famous banker and economist Muhammad Yunus and his Grameen Bank.[17] Small-business loans by the bank to poor but determined entrepreneurs produce an income stream that can in turn finance new loans for others.

Among the many accomplishments of the Grameen Bank is a very high success rate. To date, the repayment history of the bank's mostly poor borrowers has been 99 percent, a record any commercial bank would kill for. One reason is that it involves groups of "neighbors" who cooperate and provide mutual support. Yunus and his bank won the Nobel Peace Prize in 2006, and the idea has since been exported to the United States, which—sadly enough—has a large unmet need. The Grameen Bank model is also being spread to other needy countries through the United Nations Foundation and a $1 billion gift from CNN founder and philanthropist Ted Turner.

A more indigenous example, but similar in spirit, is the New York City nonprofit organization Credit Where Credit Is Due. Under the slogan "creating opportunity for all New Yorkers," this unique organization is providing financial management training to some of the city's poorest workers, 40 percent of whom have significant debts and no savings, and who rely on usurious check cashing services and loan sharks rather than using traditional banks. When any of its clients are ready to take on the challenges of saving and starting a business of their own, Credit Where Credit Is Due steers them to its affiliated Neighborhood Trust Federal Credit Union for financial support and advice.[18]

Another example is the Clinton Global Initiative (CGI), led by former president Bill Clinton and backed by a global who's who with deep pockets. Although it is far more diversified in its purpose and more global in scope, CGI has also become an important catalyst for grassroots economic and social change in various countries (like Haiti). Among the many projects the CGI has undertaken in recent years is the Harlem Small Business Initiative, a proving ground for ways to help inner-city entrepreneurs, including hundreds of small business ventures in New York City's deeply depressed Harlem area.[19]

These and other innovative efforts offer hope that there may be a way to lift many of the poor in our society (and others) out of the poverty trap. However, even at its best, neither stakeholder capitalism nor a vibrant not-for-profit sector is capable of eliminating poverty or buffering us against future capitalist storm fronts. It is long past time for us to accept this reality and reject the self-serving ideology of the invisible hand. Not even Adam Smith believed free markets were an all-purpose solution.

To paraphrase the philosopher George Santayana's long-ago warning, having forgotten the lessons of the Great Depression, we are now being forced to relive them. We need a government that augments, complements, and polices the private sector. We need a government that can act on behalf of the public interest and the general welfare, as social democrats (following Plato) have long insisted.

A compelling example (and symbol) of how our government can act as a steward of the common good and serve the interests of future generations is our unique chain of national parks, national monuments, historical sites, national seashores, and more, numbering 391 all told and encompassing 84 million acres. As filmmaker Ken Burns showed in his breathtaking documentary series on the history of our park system, which he subtitled *America's Best Idea,* every one of these very special public places represents a victory in what was often a protracted struggle against the forces of greed and private exploitation. Nor has this war been forever won; in many cases economic predators of various kinds still lurk just offstage.

The vital role of government has been underscored and discussed in no fewer than six recent books by leading economists and social scientists: Galbraith's *The Predator State,* Reinhart and Rogoff's *This Time Is Different,*

Akerlof and Shiller's *Animal Spirits,* Krugman's *The Return of Depression Economics and the Crisis of 2008,* Reich's *Supercapitalism,* and Stiglitz's *Freefall.* Akerlof and Shiller sum up the argument this way:

> *Left to their own devices,* capitalist economies will pursue excess, as current times bear witness. There will be *manias.* The manias will be followed by *panics.* There will be joblessness. People will consume too much and save too little. Minorities will be mistreated and will suffer. House prices, stock prices, and even the price of oil will boom and bust. . . . [Government] should give full rein to the creativity of capitalism. But it should also countervail the excesses that occur because of [our] animal spirits.[20]

Sometimes an entire economy becomes an "unnatural disaster," and at such times (contrary to Ronald Reagan's perverse propaganda) government is part of the solution.

The third major element of the Fair Society model—the third leg of the stool, so to speak—relates to the precept of reciprocity. The question is, How do we close the loop? How do we offset the many benefits of a biosocialist program with appropriate contributions? What social obligations should we as the beneficiaries undertake in return? Part of the answer is to be responsible, productive, law-abiding, taxpaying citizens, of course. But I believe we should expect much more. Call it a public service ethic.

National service in the form of a military draft was an obligation that previous generations of Americans accepted as a legitimate duty, although the tragedy of the much-hated Vietnam War undermined this tradition. Nowadays, national service programs like the Peace Corps and Americorps provide other opportunities for public service beyond the military, and more new service programs have recently been added. In response to a call from President Obama, Congress early in 2009 passed the Edward M. Kennedy Serve America Act, which provides $6 billion over five years to create four new service "corps" in education, health care, energy independence, and aid to veterans, as well as a "Summer of Service" program for middle- and high-school students.

But this is only a start. Why not expand on these precedents and establish a lifelong public service obligation? This would include, say, one full year of national service for everyone who is able to do it, or two years for those who receive special benefits like college tuition assistance. In addition, we could build on our rich volunteer heritage as a nation and greatly augment the volunteer work that is already being done by so many of our churches and civic organizations. Everyone should be expected to volunteer for ongoing unpaid community service—like the firefighters and hospital aides who still serve in many of our communities. If all of us were to volunteer time regularly, imagine how our local communities would be enriched. We should take to heart the saying attributed to the great fourteenth-century Muslim scholar and polymath Ibn Khaldun: "He who takes from society without giving back is a thief."

More controversial is the issue of how all the benefits associated with a basic needs guarantee program would be paid for. The answer, in brief, is that many changes will be required in the interest of paying our fair share of the costs. One important change would be a more equitable tax system. The existing system is unfair in so many ways that even tax experts probably couldn't identify them all. This is not about "soaking the rich" but rather about paying and receiving a fair share (and eliminating unjustified tax incentives).

Among the many reforms that would help to rebalance the scale of tax justice (assuming we could mobilize the political will) would be to eliminate property tax deductions for second (vacation) homes, tax capital gains at the same graduated rate as earned income, and eliminate the expanded home equity line of credit loan provisions—originally intended mainly for making home improvements—that nowadays allow homeowners to use these loans as low-interest credit cards and even to deduct the interest payments from their taxes (up to a $100,000 loan balance). We might also impose "luxury" sales taxes on high-end items (as we used to do)—like mega-mansions, airplanes, yachts, expensive fur coats, trophy cars, and collectibles. Perhaps, too, we might give tax credits to renters to offset homeowners' property tax deductions. Then there is the continuing scandal of unjustified and market-distorting farm subsidies, oil and gas subsidies, and others.

Beyond this lies the third rail—higher income taxes, especially for the superrich (as our currently weak economy allows). Conservative outrage over the idea of a tax hike is, in fact, a smokescreen designed to hide the truth. To repeat, we have the lowest overall average tax burden of any of the industrialized nations (currently at 26.9 percent of GDP, according to the OECD). Several years ago Matthew Miller, a former White House budget official, in *The Two Percent Solution,* developed the thesis that if we as a nation devoted just 2 percent more of our GDP to social welfare, it could effectively pay for all our unmet social needs. One can quibble with his arithmetic, but the fundamental point remains true: a basic needs guarantee is "affordable" at the margin in our economy, though it would certainly require a modest pay cut for the wealthiest members of our society. As Miller pointed out, we can't solve our deep social problems without money, but as a society we do have the money. "Either we tackle these challenges together, or we go on pretending to solve them while letting them fester until they explode down the road."[21]

Another measure that could also make a significant difference would be to seek greater efficiency in our social services. Thus, if we were ultimately to adopt a "single-payer" health insurance system (similar to Medicare) instead of continuing to subsidize the profits and administrative (and lobbying) expenses of the private health insurance industry, we could probably at a stroke save 20 percent on our national health insurance costs (or about $150 billion a year) and even more if various operating efficiencies were also achieved. (Again, many other industrialized countries spend less than half as much as we do and get better results.)

A Fair Society program raises many other contentious issues, but I will comment here on just a few of them. One has to do with how we can overcome the "we/they" duality in human nature that I discussed earlier—our predisposition toward moral and emotional disengagement that encourages personal and corporate selfishness, social discrimination, and social conflict. This deep-seated psychological propensity is a major culprit in the struggle for fairness and social justice in any society—and between societies as well.

There are no easy answers to this problem, to be sure, but we can collectively choose to work toward overcoming this innate bias. This is how America's blighted history of slavery and racial discrimination was finally confronted and is gradually being overcome. In other words, we can exert collective self-control over the innate impulses that support injustice and make changes in what are considered acceptable norms. The evolution of gay rights in this country is an especially important current example. We can also make conscious efforts, both as individuals and as a society, to treat everyone with respect and focus on what we have in common rather than dwelling on our differences. Moreover, when there are legitimate conflicting fairness claims, we should resist the rush to anger, hatred, and demonizing in favor of negotiation and compromise wherever possible. This is what it means to be a civilized society.

Sometimes, of course, a compromise solution is not possible, but we will never know if we don't try for it. Nor should honest conflicts of interest be allowed to degenerate into shouting matches, or violence and bloodshed. It's also important to remember that every new generation needs to learn tolerance and civility, because the evidence shows that we are born with socially polarizing predispositions. By the same token, we must in all honesty be conscious of our individual differences, which make social learning all the more challenging for some of us, especially when it comes to fairness.

Another difficult issue relates to how a fair society can survive in an intensely competitive and often exploitative global economy. When any nation gains an advantage by callously exploiting its workers, it is unfair both to them and to the workers in other countries who may lose their jobs in any economic "race to the bottom." By the same token, I noted earlier how the financial leverage of the rich countries and multinational corporations has been used in recent years, especially under the World Trade Organization and the so-called Washington Consensus, to cause "harm" to other nations and their workers. This too is deeply unfair.

The latest version of this exploitative pattern is what has been labeled "agro-imperialism." Wealthy food-importing nations, especially in the Middle East and China, have recently been buying up huge tracts of arable land in Africa (hundreds of thousands of acres at a time) and developing equally huge export-oriented farms to ship food back to their homelands. The extent of this new trend and its ultimate consequences for the indigenous African populations are still unclear. Thus far, the foreign investors have shown little interest in the locals except as low-wage laborers. But various parts of Africa have been subject to severe famines in recent decades, and hunger is always a threat. So this represents a very ominous trend.

These modern-day examples of imperialism highlight the fact that economic fairness is also a profoundly important global issue. If we hope ultimately to achieve a stable international system, it must be based on what might be called a global biosocial contract. Each nation has a responsibility for social justice, both within its borders and in its relationships with other nations. The same three fairness precepts (basic needs equality, due regard for merit, and reciprocity) are moral universals. If genocide and other horrific acts are crimes against humanity, then (to a lesser degree but similar in nature) so are the crimes that permit ostentatious islands of great wealth surrounded by deep oceans of poverty, disease, and human suffering. No religion condones it, nor does the Golden Rule, nor do our deepest moral sensibilities—if we listen to their "whisperings." It is time to make a fair global society our common cause as a species.

This goal has implications for the larger international political system, where the traditional paradigm has been based on international rivalries and a struggle for power. The power politics model is increasingly outdated, and it was never concerned with social justice to begin with. It was based on a competitive pursuit of national self-interest, and with very few exceptions it was also a highly exploitative system. But as the game theorists teach us, such a system is inherently unstable and must ultimately rely on coercion (military force), or the threat of it. Nowadays our growing economic interdependency and the global scope of our problems (climate change and international terrorism being only the most obvious examples) require stable, cooperative solutions, which will be possible in a sustained way only if based on the fairness precepts.

The Harappan/Indus civilization, described in chapter 3, may be a good model for our global future. Recall how archaeologist Charles Maisels characterized this diversified, sprawling network of villages, towns, and cities as a "commonwealth"—a self-organized, voluntaristic system. It was based on mutually beneficial trade. It avoided the extremes of wealth and poverty that arose in every other ancient (and modern) civilization. And all the evidence suggests it enjoyed relative peace. The world is a much more crowded and complex environment today, and there are many challenges and stresses that severely test anyone's resolve to act fairly. But the goal is still valid, and we should always keep the target in our crosshairs, even if at present it remains out of range.

Recent calls for the globalization of "empathy" and concern for people in other nations are certainly an encouraging sign, and the outpouring of help for the Haitians in the wake of their catastrophic earthquake was one heartening indicator—perhaps one of our finest hours as a global community. Many years ago, the philosopher Peter Singer coined a term for this hopeful trend. He called it "the expanding circle."[22]

Another serious challenge to achieving a Fair Society, and a fair global society as well, is the ominous rise in economic and social stress, which is destined to become much worse given the combination of continued population growth, declining natural resources, and a relentless increase in global warming—with all it implies for disrupting global food production. Increasing scarcities are likely to engender deep social conflict in many hard-pressed societies. Agro-imperialism may be only a bellwether of the future. (The scarcity issue is discussed further in the endnote.)[23]

As we witnessed in ancient Athens (chapter 3) and have experienced many more times over the past two thousand years, a society under severe stress can abandon any semblance of fair dealing and social justice and can come to resemble a Hobbesian war of "each against all." (Remember the Ik in Colin Turnbull's study?) A disturbing foretaste of this syndrome is the rise of vitriolic hatemongering in our own deeply stressed society by right-wing radio and TV demagogues, who have been promoting virulently antisocial behavior among a well-armed and increasingly paranoid minority.

Witness the fulminations of Michael Savage (aka Michael Weiner) in his provocatively titled book *The Savage Nation: Saving America from the Liberal Assault on Our Borders, Language, and Culture*. Savage/Weiner tells us that he speaks "the truth" about "the terrorist network operating within our own borders." Who are they? "Every rotten, radical left-winger in this country, that's who."[24] This profoundly un-American conduct is symptomatic of a serious social pathology, and it is a warning to us that much worse might come of it if we cannot find the political will to rewrite our social contract.

The evolutionary psychologist Dennis Krebs summed it up this way: "To create moral societies, we must make it in people's adaptive interest to cooperate with others, and the only way to do this is to design environments in ways that ensure that cooperation pays off better than selfishness, cheating, free riding and favoritism."[25] To ensure voluntary cooperation and social harmony, in other words, we must first achieve social justice, plain and simple.

However, there is a daunting political obstacle that stands in the way of achieving a Fair Society in this country, or anywhere else for that matter. How do the roughly 70 percent of us who support the principle of fairness and social justice overcome the formidable power of the 30 percent who largely control our politics and our wealth and who will fiercely defend the existing system, and their self-interest? Some captains of industry are scornful of the very concept of social justice.

True, there are also many wealthy philanthropists and enlightened billionaires, like Bill Gates, Warren Buffet, Ted Turner, and even Pierre Omidyar, who are willing to do their fair share for our society. But this is obviously not enough. What is needed is a collective (democratic) decision by the 70 percent who are fair-minded to impose the needed changes on the retrograde 30 percent, rather than allowing them to dictate the rules of the game—and the outcomes. In a very real sense, we must restore the "reverse dominance hierarchy" (see chapter 3) that our wise hunter-gatherer ancestors devised to constrain the egoists, power seekers, and bullies that have been a feature of human societies for many thousands of generations.

History teaches that every society with a mixed democratic government has the power—if we choose to exercise it—to create a fair society through the political process. This is precisely what those European welfare capitalist societies have done, and so can we. All the others, especially the exploitative authoritarian societies, may be destined to relive one of the other great lessons of history—to suffer the civil conflicts that everyone from Plato to Hobbes, Adam Smith, Karl Marx, and Barrington Moore have warned us about. Capitalism and socialism both promised material progress and a better life. Both have ultimately disappointed us. The Fair Society promises only social justice, whatever the future may bring. But that will be good enough.

In 1944, at the height of World War II, Franklin Roosevelt proposed a new "Economic Bill of Rights" as part of his vision for the postwar world. It included many of the basic needs documented in chapter 5. Now the science of human nature, and especially the science of fairness, has provided compelling evidence in support of Roosevelt's moral instincts. We must treat our basic needs as a social right, while not abandoning the other fairness precepts I have identified. If we deny these biological imperatives or fail to act on them, we do so at our peril.

Conservatives with vested interests in the status quo will no doubt dismiss the idea of a Fair Society as just another utopian scheme. Far from it. The naysayers need to be reminded that there is an important historical precedent they would like to forget or discount—the New Deal in the 1930s. Whatever its shortcomings, the New Deal set a new course for this country that has had enduring benefits. In effect, we need a new New Deal. Nor is the ultimate goal of a Fair Society some kind of idealistic fantasy. We are, in fact, talking about achieving a society that more closely resembles what already exists in countries like Denmark, the Netherlands, and Sweden, where, you will recall, fairness goes under the heading of welfare capitalism. As epidemiologists Richard Wilkinson and Kate Pickett show in their important 2009 book *The Spirit Level*, more egalitarian societies generally do better by many measures of health and well-being.[26]

The recently enacted health reform legislation is an important step in this direction. Among other things, it established for the first time in this country the core principle that basic health care should be a social right. It is, recall, one of our fourteen basic needs domains. As Congressman Patrick Kennedy (son of the late Senator Kennedy) expressed it, "Health care is not only a civil right, it's a moral issue."

The vision of a Fair Society presented here is, of course, only a sketch, not a detailed plan. It is meant as a goad to collective action. If conservative politics got us into this mess, only more progressive politics will get us out of it. Even so, there will no doubt be many vocal—even hysterical—opponents. Some of those conservative hate-mongers on talk radio and TV will denounce it as socialistic, totalitarian, anti-American, and so forth. But these egotistical, self-serving tirades—often accompanied by outright lies—are mean-spirited and dangerously divisive. These merchants of hatred are preaching to the devil's choir. One example, among many others, was Fox News host Glenn Beck's recent frontal attack against social justice, which he denounced as a "perversion of the Gospel." The idea is associated with communism and fascism, he claimed.[27] It is the moral equivalent of shouting fire in a crowded theater, to borrow a famous legal precedent.

More moderate opponents of a Fair Society will simply say that it is totally "unrealistic." But these cynics are like the "usual suspects" in the classic movie *Casablanca*. It has hardly been an unrealistic goal for many other advanced nations, and we do, after all, still have the capacity to make changes in our society. It has historically been one of our strengths as a nation. We have already made a start with the changes to our dysfunctional health care system.

But how do we get from here to there? Where do we begin? I'm reminded of an inspiring World War II incident. On D-Day, June 6, 1944, the leading elements of the U.S. Fourth Infantry Division went ashore at Normandy far from their designated landing site. When the assistant division commander, Brigadier General Theodore Roosevelt Jr. (son of our twenty-sixth president), who was huddled on the beach with his men, realized there had been a serious landing error, he ordered his division to advance anyway with the famous remark, "We'll start the war from here."

We, too, must start the war from here. As I noted earlier, major political changes require both political leadership and political mobilization. We need the functional equivalent of a Martin Luther King Jr. or an Al Gore, a skillful champion dedicated to the overarching goal of achieving a Fair Society. He, or she, or they will also need many divisions of foot soldiers. However, I am not proposing the creation of some new political reform party. Whatever may be its political and programmatic shortcomings, the Green Party already occupies the reformist niche in our politics. Moreover, trying to compete for political power as a third party may not be productive. It has been tried many times in our history without success.

Instead, what is most practical, I believe, is to build a nonpartisan political movement—a "fairness coalition"—that can act in a concerted way in every state and every congressional district to change the hearts and minds (and the campaign funding sources) of our political establishment and enable (or compel) our political leaders to act on meaningful reform measures even in the face of fierce resistance by the coalition of the unwilling—the vested interests that still control our damaged economic system and dominate our politics. In a recent op-ed column in the *New York Times,* Paul Krugman concluded: "Turning this country around is going to take years of siege warfare against deeply entrenched interests, defending a deeply dysfunctional political system."

At this traumatic time in our history, we have come to a major fork in the road. To quote one of the sayings of America's favorite populist sage, the baseball player and manager Yogi Berra: "If you come to a fork in the road, take it."[28] In fact, we have no choice but to take one of the alternative paths laid before us, even if we fail to exercise a deliberate choice. We can either choose to take the high road toward a Fair Society, or we can continue down a much darker road where the shadows are deepening and great dangers lurk. If we abdicate this responsibility, we will then leave the choice to others who do not share our goals.

One of Martin Luther King Jr.'s most inspiring lines is, "The arc of the moral universe is long, but it bends toward justice." President Obama has observed that it does not bend by itself. Nor, in fact, can any leader bend it by himself or herself. It is up to us, the 70 percent who are committed to fairness, to impose our collective will on the other 30 percent who are fairness "challenged." As Plato and Aristotle tried to tell us, the greatest good for the greatest number is, after all, a fair society. Fairness, in the end, is all about what we do (or don't do) for each other. If we fail to respond to the wisdom of the ages, as Goethe admonished us to do, we will surely have the war Thomas Hobbes warned us about in the epigraph for this chapter. How long will it take us to learn this lesson?

Epilogue: What Can I Do?

The whole is greater than the sum of its parts.
PARAPHRASE OF ARISTOTLE IN THE *Metaphysics*

The only answer to organized money is organized people.
BILL MOYERS

The natural world is chock full of amazing collective feats—the orb web spiders that combine their efforts to build huge webs that can span a mountain stream, the raiding parties of army ants, numbering 200,000 or more, that can bring down and dismember a lizard, a snake, or a nestling bird, the tropical bee colonies that form temperature-controlled nests for their young out of tens of thousands of their interlinked bodies, or the ten trillion or so highly specialized cells that make up a human body. Talk about a combination of labor! And think of all the things humankind has accomplished collectively, from the pyramids in Egypt to the Great Wall of China, from our awe-inspiring cathedrals and mosques to the moon walks by the *Apollo* astronauts, thanks to the combined efforts of 400,000 NASA employees, innumerable industrial contractors, and a supportive nation.

In the end, a fair society can be achieved only by collective action. This too is one of the great lessons of history. No individual, not even the most exalted and inspiring leader, can be anything more than an instrument for accomplishing our common goals. And this is especially true of a task as huge and challenging as rewriting our social contract. But if we make it our common cause to create a fair society, then our individual contributions can be leveraged by the contributions of others. This is the take-home lesson from the presidential campaign in 2008, which broke all records for volunteer participation and relied on the many millions of small contributions from individual donors. Collectively, these actions helped determine the outcome.

So there are, in fact, many things you can do to help bend the arc of the moral universe. For starters, you could share this book with others and encourage them to read it. You could join in the discussion at our blog, TheFairSociety.blogspot.com, or contribute to the work on a Fair

Society agenda at our Web site: www.TheFairSociety.net. Or perhaps you could start, or join, a discussion group or a blog focused on one or more specific fairness issues at any of three levels—local, national, and international. Feel free also to share your work with us at our e-mail address: TheFairSociety@gmail.com.

Beyond this, you could join an organization working on fairness issues that are of special concern to you. You could also do volunteer work for candidates or elected officials who are strongly committed to achieving a fair society, especially those with an unqualified commitment to ensuring that the basic needs of all of our people are adequately provided for. Not least, you could contribute whatever you can financially to help those organizations and candidates for public office who will make a difference.

Perhaps most important of all, you could make a personal commitment to think and act fairly toward others yourself. To quote Bill Moyers (again) in one of his recent PBS TV programs, "Fairness is a choice to be made, a responsibility to be honored." Sometimes this can be hard, and costly, and we must always be on guard against the outliers who will gleefully exploit anyone who is trusting or who tries to be fair—not to mention the subversive influence of our self-serving impulses. Nevertheless, the single most important finding of the multidisciplinary science of fairness is that most of us do have a bias toward cooperation and a readiness to reciprocate—a sense of fairness. A Fair Society has the legitimacy of science behind it: it is not a fantasy based on some simplistic view of human nature or some outworn nineteenth-century ideology.

So if you make fairness your personal modus operandi, it is likely that it will ripple outward and affect the behavior of others, like a pebble thrown into a pond. As the longshoreman philosopher Eric Hoffer (author of *The True Believer,* among other works) put it many years ago, "Kindness is not a role you play; it's a chain reaction." If each one of us tries always to act fairly and insists that others do so as well, this could go far toward producing the changes we need.

Notes

PREFACE

1. Organisation for Economic Co-operation and Development OECD statistics for 2007: http://www.oecd.org/dataoecd/48/56/41494435.pdf.

2. From a study by Jared Bernstein, Brocht, and Spade-Aguilar 2002. See also Galbraith 1998, Corning 2003b.3. Nord, Andrews, and Carlson 2009.

3. Nord, Andrews, and Carlson 2009.

4. This was how economist James K. Galbraith characterized it in *The Predator State* (2008). See also the *New York Times* column by Paul Krugman, "Looters in Loafers," http:www.nytimes.com/2010/04/19krugman.html.

5. Among others, see Paul Krugman, "How Did Economists Get It So Wrong?" http://www.nytimes.com/2009/09/06/magazine/06Economic-t.html?page wanted=all.

6. http://www.oecd.org/document/30/0,3343,en_2649_34631_12968734_1_1 _1_37407,00.html.

7. Near the end of 2009 it stood at 46.3 million, according to the Progress Report from the Center for American Progress for September 16. http://www.progress @American progress.org.

8. http://www.cia.gov/library/publications/the-world-factbook/rankorder/2091 rank.html, and http://www.cia.gov/library/publications/the-world-factbook/rank order/2102rank.html.

9. The estimate is from a study by the American Society of Civil Engineers. http:// www.asce-sf.org/index.php?option=com_content&task=view&id=505&Itemid =42.

10. Quoted in Frank Rich, "Hollywood's Brilliant Coda to America's Dark Year," *New York Times*, December 13, 2009.

11. In an interview on PBS *News Hour*, February 2, 2010.

12. Lippmann 1989.

13. Smith 1964.

14. Smith 1976.
15. Reich 2007, 13.

CHAPTER ONE: LIFE IS UNFAIR

1. http://www.supremecourtus.gov/opinions/06pdf/05–1074.pdf. Also "Congress Relaxes Rules on Suits over Pay Inequity," *New York Times,* January 27, 2009.

2. http://www.rollingstone.com/news/story/10820245/people of the year 2004 barack obama.

3. http://www.aclu.org/womens-rights/aclu-cheers-house-passage-pay-equity-legislation.

4. Gouldner 1960. See also Westermarck 1971, MacCormack 1976, Salter 2008.

5. Among others, see Plato 1946, Aristotle 1946, 1985, Cicero 2001. For in-depth studies, see Barker 1960, Harris 2006.

6. Rawls 1995, 1999, Walzer 1983, Masters 1989, Masters and Gruter 1992, Klosko 1992, Wilson 1993, Joyce 2007. Also, see Cohen 1986, Price and Feinman 1995, Wolff 1996, Ridley 1997, Arnhart 1998, Katz 2000.

7. Rousseau 2004.

8. Sober and Wilson 1998. For a book-length study of group selection and its implications for human altruism, see Field 2001.

9. A chapter-length history and discussion of this concept can be found in Corning 2005.

10. Maslow 1962.

11. Doyal and Gough 1991.

12. See the full-length discussion in Corning 2005.

13. See Gintis et al. 2005.

14. Two other important uses of the term "fair society" should be mentioned here. British prime minister Gordon Brown adopted it in the fall of 2008 as a political slogan and campaign theme for his reelection bid, and sociologist Amitai Etzioni has espoused the concept in connection with his research on fairness. Although there is common agreement about the fundamental principle, Brown and Etzioni used the terms in different ways. Brown's rhetoric was impressive. He called fairness "a cause worth fighting for" and espoused "a fair Britain for a new age." Why fairness? "We do it because fairness is in our DNA. It's who we [the Labour Party] are. . . . Labour stands for a fair society." However, Brown provided no theoretical basis for his vision. It served as the packaging for a set of incremental policy proposals, from free nursery care for two-year-olds to vouchers to allow low-income citizens to gain access to the Internet. Indeed, it seemed the term did not poll well as a campaign slogan, because it was later dropped in favor of "building a better Britain." Etzioni's vision is more ethically grounded and less programmatic. "We must work together for a fair society: a society where everyone is treated with full respect, recognizing that we are all God's children." An accompanying survey—conducted for him by a professional survey research firm in 2004—provided impressive support for his argument that the norm of fairness has strong backing across the political

spectrum. In survey questions related to the fairness of a variety of economic and political issues, "unfair" responses ranged from 52 percent for privacy violations in the Patriot Act to 88 percent for health insurance companies' denials of medical cost claims by their customers. For Brown, see www.independent.co.uk/news/uk/politics/brown-fights-back-with-fair-society-vision-939767.html and www.youtube.com/watch?v=49L6RHuknsw. For Etzioni, see his chapter "The Fair Society" in Etzioni 2005.

CHAPTER TWO: THE IDEA OF FAIRNESS

1. White House Press Conference, March 21, 1962.

2. Of course, if you are a person of religious conviction you may see things differently. If you believe in an omniscient, omnipotent supreme being, you might view life in terms of a cosmic system of rewards and punishments. You might also adhere to the view of Saint Augustine that, while there can be no justice on earth, you will receive "divine justice" for better or worse in the hereafter. This is the inspiration for the Christian belief in heaven and hell. For Christians, who see injustice all around them just as the rest of us do, and who are often themselves the victims, this is an important source of consolation harking back to the teachings of Jesus. Of course, this belief system has also provoked an enduring theological question, Whence cometh evil? Or to paraphrase the title of a popular book of a few years back, Why do bad things happen to good people? For some it all dates back to Adam and Eve and "original sin." A similar theme can be found in Islam. Among the many modern-day discussions of fairness and social justice, see especially Rawls 1999, Miller 1976, 1999, Pettit 1980, Frank 1988, McShea 1990, Wilson 1993, Masters and Gruter 1992, Young 1994, Kolm 1996, Solomon and Murphy 2000, Raphael 2001, Akerlof and Shiller 2009.

3. Milgram 1974.

4. Elster 1992, 1995.

5. Young 1994.

6. See McCullough 1993.

7. Peyton Young 1994, in his important book on the subject of equity, provides this wry preface: "When I teach [equity] to students, I warn them that the subject does not exist. Among all nonexistent subjects, in fact, equity occupies a distinguished position because it fails to exist in several different ways. The arguments against existence take three different forms. The first is that equity is merely a word that hypocritical people use to cloak self-interest. . . . The second argument is that, even if equity does exist in some notional sense, it is so hopelessly subjective that it cannot be analyzed scientifically. . . . The third argument is that, even granting that equity may not be entirely subjective, there is no sensible theory about it. . . . In short, it fails to exist in an *academic* sense. . . . Set against these arguments is the fact that people who are not acquainted with them insist on using the word 'equity' as if it did mean something. In everyday conversation, we discuss with seeming abandon the equities and inequities of the tax structure, the health care system, the military

draft, the price of telephone service, and how offices are allocated at work. For a term that does not exist, this is a pretty good beginning" (x). See also Walster, Walster, and Berscheid 1978, Deutsch 1985, Brams and Taylor 1996, Miller 1976, 1999, and Spector 2008.

8. Quoted in Gouldner 1960. See also Cicero 2001.

9. Among the many discussions, see especially McShea 1990, Wattles 1996, Solomon and Murphy 2000, Raphael 2001.

10. A colleague, Roger Masters, points out that the Old Testament is actually inconsistent. Deuteronomy 19:21 commands revenge: "It shall be life for life, eye for eye, tooth for tooth, hand for hand, foot for foot."

11. Kant 1997, 4:404. See also Kant 1999.

12. Binmore 2005, ix. Jeffrey Wattles, in *The Golden Rule* 1996 is among those who side with Kant. He claims that the categorical imperative differs because it is focused on principles of our "will" and not "actions" to be done to others or ourselves; it focuses on "rules"; and it requires rational judgment about what is appropriate as a universal law. "Reason must validate them" (83). To me, this is philosophical hairsplitting. If the Golden Rule doesn't qualify as a categorical imperative, so much the worse for the categorical imperative.

13. Rawls 1999.

14. Binmore 2005, 129. Binmore also notes that the economist John Harsanyi many years ago independently developed a formulation that was similar to Rawls's, though it is not so well known. Some critics have pointed out that it makes more sense logically to opt for economic equality. And in fact Rawls did just that in an earlier article, but he shifted to the "least favored" in his famous book. Others have questioned whether a hypothetical situation with no relation to the real world—comparable to an unrealistic thought experiment in science—can legitimately be used to derive "intuitive" principles for real world application. Still others object that Rawls's two principles seem potentially to produce self-contradictions. On one hand, if economic inequalities are allowed to persist, this would constrain the equality of others—the purchasing power and freedom from want of the have-nots. On the other hand, imposing constraints on allowing the rich to be able to benefit from the fruits of their economic accomplishments limits their freedom to hold property. We will come back to these issues later on. Among the critical discussions, see especially Wolff 1996, Alejandro 1998, Raphael 2001.

My own criticisms are more fundamental. Rawls recognized what he called "primary goods," which he defined as "rights and liberties, opportunities and powers, income and wealth," but his theory did not address basic needs per se or give primacy to their satisfaction; basic needs did not seem to rise to the level of a moral imperative for Rawls. Instead, his principle would ensure only that the poor get a piece of the action when the rich get richer. It amounts to a pledge that a rising tide should lift all boats. In a later book, Rawls shifted the rank ordering of his normative principles and placed basic needs first, above liberty. But his rationale for doing so was curious. He argued that it was justified because meeting basic needs was a prerequisite for freedom. Another criticism I have is that Rawls tolerated inequalities, yet he did not make any explicit provision for "merit"—rewards for talent, effort, and achievement. As various critics have noted, Rawls was not much concerned about fairness as "just deserts." Nor did he address the free rider problem. See also the discussion in Corning 2005.

15. http://en.wikipedia.org/wiki/The_Golden_Rule_ethics. Some of these objections border on the absurd. For instance, what about a sadomasochist who ventures forth to do unto others . . . ? Then there is the condescending view of the twentieth-century theologian Paul Tillich, who considered the Golden Rule "inferior" to unconditional love—that is, altruism. Of course the good Samaritan, the all-time bestselling Golden Rule role model, did act altruistically, as we would want others to do when it's appropriate.

16. In fact, there are now "postconviction" statutes in forty-eight states that allow DNA testing and the release of wrongly convicted felons. At this time, some 250 innocent prisoners have been released and, in some cases, compensated by the state.

CHAPTER THREE: A BRIEF HISTORY OF (UN)FAIRNESS

1. Dawkins 1989, v.

2. Nozick 1974, ix.

3. Rand 1993, 678–83. See also the recent assessment of Rand in Burns 2009. It is worth noting that Ayn Rand's moral stance in *The Fountainhead* is actually subversive. In effect, she sets the rights of her creative individuals above the law and above the legally recognized property rights of others, including especially those "second raters," "parasites," and repulsive "masses" that she often denigrates. In the story, Roark agrees to secretly help a mediocre old school friend, Peter Keating, win a large housing development contract on condition that there must be no changes at all to Roark's innovative plans. Keating agrees and, in fronting for Roark, dutifully inserts a similar "no change" clause into his contract with the client/owner. But as the buildings go up, the owner violates the contract by unilaterally making some cosmetic changes. So Roark sneaks onto the project site one night and blows up all the buildings. Then, after he confesses to the act but makes a long-winded philosophical justification to the court, excerpted above, the jury acquits him! This is perverse. Roark's grievance was with his friend. He had no contract with the owner whose property he destroyed, and no legal claim against him. It was Keating who failed to insist on and defend the "no change" clause. Roark's recourse was to bring a lawsuit against Keating. But Roark was not bound to act in accordance with the law, or the principle of punishing only the culprit, or even respecting property rights. And neither, it seems, were the judge and the jury! This is frankly absurd, even as fiction.

4. http://www.usatoday.com/money/companies/management/2002-09-23-ayn-rand_x.htm.

5. http://online.wsj.com/article/SB123146363567166677.html. See also Rand 1957.

6. Rubin 2002, xiv.

7. This narrative is a synthesis of incidents described by Kummer 1968, Strum

1987, and Van Hooff 2001. Collective decision making has also been observed in other social species, including red deer, gorillas, chimpanzees, and African buffalo.

8. Strum 1987.

9. The following description is drawn from the more detailed and documented scenario for human evolution in Corning 2003a. Also relevant are the extensive book-length treatments by Klein 1999, Wolpoff 1999, and Ehrlich 2000.

10. Wolpoff 1999.

11. Kingdon 1993.

12. Boehm 1999.

13. Shultziner et al. 2010.

14. Maryanski and Turner 1992. In their original rendering of the term social cage, Maryanski and Turner used it in a pejorative sense as a set of unnatural social constraints that violate our "primate natures." They believed that humans are not particularly social "at a biological level" (vii). Primatologist Frans de Waal 1996, on the other hand, uses the term in a more positive way, in line with the overwhelming evidence that we and our primate relatives are in fact highly social animals. Yet we are also highly egoistic and self-serving. So both versions of the term have some merit, I would argue.

15. Gowdy 1998, xv–xvi. Anthropologist James Woodburn's classic 1982 paper "Egalitarian Societies," which emphasized the transition in our economic evolution between a system of "immediate returns" and "delayed returns," also remains a useful contribution to our understanding of hunter-gatherers.

16. Steward 1938, 1941.

17. Darwin 1874.

18. See Corning 2007 and the many references there. Also, see Keeley 1996 and Gat 2006.

19. Gladwell 2002.

20. Diamond 1997.

21. Alexander 1987.

22. Carneiro 1970. An example of this dynamic can be found in the history of the Zulu nation discussed in Corning 2005, 2007. Also, see Gluckman 1940, 1969, and Morris 1965. See Corning 2005, 204.

23. See Corning 2007.

24. Why is it that some quite warlike societies—like the Yanomamö of Venezuela or the Dani of New Guinea—did not evolve into nation-states? Why did some societies achieve statehood and then subsequently collapse or even disappear? And why did the first pristine states appear during a very small slice of time in the broader epic of evolution, within a few thousand years of one another at most? Finally, there are the cases in which population pressures were relieved by increased trade or by an intensification of subsistence technologies. A prior question is why populations are able to grow in some circumstances and not others. It is not a given. In fact, there are numerous cases in which chiefdoms and states failed to emerge from a circumscribed context—when the environmental vise was in fact too tight for fur-

ther expansion. Nor is warfare always correlated with population pressures. There is even some evidence of cases where population pressures and warfare were the result of economic and political integration rather than the reverse. In sum, warfare may have had an important influence in human history, but it is neither necessary nor sufficient to account for the evolution of modern societies.

25. Pfeiffer 1977.

26. Fagan 1998.

27. Trigger 2003, 375.

28. Maisels 1999.

29. Ewald 1991.

30. http://www.economicexpert.com/a/Gregory:Bateson.htm.

31. *Federalist,* Madison 10. A full text can be found online at www.foundingfathers .info/documents.

32. The following discussion is based on Connolly 1998, Plutarch 2001, Boardman, Griffin, and Murray 2002, Waterfield 2004, Harris 2006, and Pomeroy et al. 2007. Note also that, even in its golden age, Athens was a bundle of contradictions and paradoxes that deeply compromised the fairness principle. One of the most striking was that only native-born citizens could vote, or about 40,000 out of a population estimated to be 300,000 or more. Women, of course, could not vote. And neither could the 80,000 slaves and 40,000 metics (resident foreigners, some going back generations). Slaves were on the whole relatively well treated and often moved up into the ranks of skilled craftsmen. Nevertheless, the large slave population helped to subsidize Athens's affluence. And if Athens was in principle democratic in its domestic politics, it was authoritarian and exploitative in its treatment of its colonies, extracting high taxes and aggressively punishing evaders. Finally, an argument can be made that democracy in Athens succeeded in part because poor citizens were well paid for attending Council meetings and sitting on juries. It was a practice that ultimately antagonized the wealthy landowners.

33. Quoted in Plutarch 2001, *Pericles,* part 1.

34. Ibid.

35. Thucydides 2004, 2.41.1.

36. Plato 1946. See also Barker 1960, Sabine 1961.

37. Plato 1992a, also 1992b.

38. Aristotle 1946, 1985.

39. Thucydides 2004, 5.89.

40. Huxley 2005, 347ff.

41. Plato 1946, 2.369a-d, 370b, c, 372a.

42. Plato 1992a, 1992b.

43. www.dictionary-quotes.com/a-government-of-laws-and-not-of-men-john-adams/-.

44. See especially the discussion in Sabine 1961, 74–78. A brief comparison between Plato and Ayn Rand might also be useful here. Both were idealists and elitists. Neither trusted the "masses"; they were certainly not democrats. Both would also

allow freedom of action for special individuals. However, the similarities end there. Plato's philosopher kings were empowered to be perfect instruments for achieving social justice. Ayn Rand, the high priestess of selfishness, advanced the cause of creative geniuses and their self-interest. Plato favored selfless rule over the masses by people who were specially educated and devoted to duty. Ayn Rand sought freedom of action for exceptional individuals and had no interest in, or concern for, society. For Plato, social justice involved giving every man his due. For Ayn Rand, the concept of social justice was incomprehensible. The only form of justice she recognized was unfettered freedom for the Howard Roarks of the world. For Plato, a society is a social organism whose purpose is to meet human needs and wants. For Ayn Rand it is simply a marketplace (assuming that an oppressive democratic government can be removed from the picture), and the freer it is the better. Indeed, in a perfectly free market, all would benefit from the free play of their own self-interest. Finally, both Plato and Ayn Rand were misguided in their idealism, but Plato came to recognize his errors and went back to the drawing board. Ayn Rand was, like Howard Roark, uncompromising to the end, despite a turbulent history of personal relationships that might have called some of her assumptions into question.

45. Aristotle 1946.

46. Aristotle 1946, 1.1253a31.

CHAPTER FOUR: FAIRNESS AND THE SCIENCE OF HUMAN NATURE

1. Durkheim 1938, 104.

2. Pinker 2002, 1.

3. Keynes 1936, 383.

4. Epicurus, Golden Maxims no. 33, quoted in Sabine 1961, 124.

5. Bentham 1970, chap. 1, sec. 1.

6. Machiavelli 1996, XVII.

7. Machiavelli 1950, I, 9, and 1996, XV. Contrary to his Darth Vader image, Machiavelli was a mild-mannered and cultured person. The scion of a line of distinguished lawyers, he became a diplomat and, on the side, was known as a talented musician, poet, and playwright as well as for writing various political tracts. Indeed, in a later and less well-known work, *The Discourses,* Machiavelli expressed a yearning for a stable, democratic society with power shared among all the classes. His vision was drawn from his reading of the early history of the Roman republic, which developed a mixed form of government with elected representatives (rather like what Plato and Aristotle had recommended). Nevertheless, what we remember about Machiavelli are lines like these: "Our religion places the supreme happiness in humility, lowliness, and a contempt for worldly objects. . . . These principles seem to me to have made men feeble, and caused them to become prey to evil minded men. . . . for the sake of gaining Paradise, [they] are more disposed to endure injuries than to avenge them" (1950, II, 2).

8. Hobbes 1962, XI.

9. Hobbes 1962, VIII.

10. Hobbes 1962, XVII.

11. Lord Acton expressed this opinion in a letter to Bishop Mandell Creighton in 1887.

12. Quoted in *New York Times,* October 28, 1973.

13. Locke 1970.

14. Libertarianism is really an umbrella term that is applied to a spectrum of political philosophies and to a bewildering variety of sects, splinter groups, and political parties around the world, some of them closely identified with one or another outsized ego. There are left and right libertarians, consequentialist libertarians, anarchocommunist libertarians, contractarian libertarians, category libertarians, Sydney libertarians, geolibertarians, monarchists, and of course Objectivists. However, the common denominator is their passionate adherence to personal liberty and the protection of individual rights, including especially private property, from the intrusions of "the state."

15. http://www.adamsmith.org/smith/tms/tms-p2-s2-c2.htm.

16. http://www.utm.edu/research/iep/b/bentham.htm.

17. This is the opening line of Rousseau's famous *Discourse on Inequality* (2004).

18. Rousseau 1984, I, ix, I, i.

19. Rousseau 1984, II, iv, I, vii.

20. Quoted in Dunn 2002, 301; also Sabine 1961, 498.

21. Among others, see especially Bellah 1991, 1995, Elshtain 1995, Etzioni 1993, 1998, 2000, 2005, MacIntyre 1981, 1988, Taylor 1992, Walzer 1983. Etzioni, for example, speaks of a "three-legged stool" where the private sector, government, and communities are coequal actors.

22. Pinker, in *The Blank Slate* (2002), identifies a fourth model of human nature, the Judeo-Christian tradition. In addition to its many metaphysical elements, there is also, in both the Old Testament and the New Testament, a dualistic view of human psychology that is rather similar to what Plato and Aristotle articulated. We are all descendants of the original sinners but are capable of redemption, the Bible says. According to recent polls, some three-quarters of all Americans more or less adhere to this vision.

23. Darwin 1874, 115–17.

24. See De Waal 1996, 1997.

25. Dawkins 1989, 2.

26. Williams 1993. See also Williams 1992.

27. Alexander 1987.

28. Darwin 1874, 146–47.

29. Williams 1966, 8.

30. Darwin 1874, 147.

31. Darwin 1874, 148.

32. De Waal 1982, 1996, 2001, 2005, 2006.

33. De Waal recounts many incidents in his various writings.

34. Flack and de Waal 2000, 6.

35. Trivers 1971 1985.

36. Wilkinson 1984, 1990.

37. Brown 1991. The recent rediscovery of the pathbreaking work on human morality by the nineteenth- and early twentieth-century Finnish sociologist Edward Westermarck, who is also credited with describing the so-called incest taboo in humans, is detailed in a recent article by human ethologist Frank Salter (2008). Among other things, Westermarck was the first to note Brown's finding regarding the near-universal existence of moral systems in human societies. Westermarck also highlighted the "ethnic particularism" or nepotism in human morality.

38. Eibl-Eibesfeldt 1989; Darwin 1965.

39. Ember 1978.

40. Divale 1973.

41. Edgerton 1992.

42. Turnbull 1961.

43. Turnbull 1972.

44. Plomin, Defries, and McClearn 1990.

45. See Bouchard 2004, Plomin and DeFries 1998.

46. Plomin, Defries, and McClearn 1990, Dilalla and Gottesman 2004, Bazzett 2008.

47. The list of physiological disorders where genes are involved is even longer, more than 120 in all. Among others, it includes phenylketonuria, hemophilia, porphyria, galactosemia, Huntington's chorea, Tay-Sachs disease, Wilson's disease, Hartnup's disease, sickle-cell anemia, albinism, congenital nephrosis, cystic fibrosis, cretinism, achrondroplasia, color blindness, and such commonplace afflictions as rheumatoid arthritis, diabetes mellitus, epilepsy, and many allergies.

48. Rushton et al. 1986. The important new science of genomics, based on the recently developed ability to decode genetic sequences directly, promises to greatly expand our understanding of the relation between our genes and our behavior in the years ahead.

49. See especially the detailed account in Damasio 1994. Also, Macmillan 2002, 2008, and Pinker 2002.

50. Damasio 1994. On moral psychology, see Haidt 2007.

51. Delcomyn 1998, Hubel and Wiesel 1979, Bear, Connors, and Paradiso 2007.

52. Gazzaniga 2005, 165.

53. Gazzaniga 2005, 171.

54. Pfaff 2007.

55. Pfaff 2007, 4.

56. *Science Daily*, September 14, 2001. http://www.sciencedaily.com/releases/2001/09/010914074303.htm.

57. Sanfey 2007.

58. Woodward and Allman 2007.

59. A definitive overview of the field can be found in *The Handbook of Evolutionary Psychology,* edited by David Buss (2005).

60. In Buss 2005, 584–626.

61. In Buss 2005, 747–71. See also the pathbreaking work of psychologists Jean Piaget (1932) and Lawrence Kohlberg (1981), and see Kohlberg, Levine, and Hewer 1983 on the development of morality in children.

62. Beinhocker 2006.

63. Keynes 1936.

64. Akerlof and Shiller 2009, ix.

65. Akerlof 2007.

66. Binmore 2005, 63. Indeed, the assumptions/specifications underlying the Prisoner's Dilemma paradigm seem, in retrospect, to be impossibly contrived. The original "story" behind the Prisoner's Dilemma involves two gangsters and a district attorney who knows they are guilty of a serious crime but needs a confession from either one of them in order to convict them. So he offers each of them separately and unknown to the other the following options:

- If you confess but your partner does not, you can go free.
- If you fail to confess but your partner does confess, you will get the maximum sentence.
- If you both confess, you will get lesser sentences.
- If neither of you confesses, you will be set up and convicted on a lesser charge.

Got that? It's no wonder most social scientists could not see how this model was broadly applicable to economic and social life.

67. Axelrod and Hamilton 1981. See also Axelrod 1984.

68. Nowak and Sigmund 1993.

69. Strong reciprocity theory, and the growing research literature associated with it, is discussed in some detail in Corning 2005, 156–57.

70. Corning 2005, 29–30.

71. In addition to the works cited in Corning 2005, 156–57, see also Gintis et al. 2005, Gintis and Fehr 2007, and Gintis 2008.

72. Henrich et al. 2001; Henrich et al. 2010.

73. Similar ideas, though without using the term, have been expressed by many theorists, from Aristotle to Herbert Spencer, Émile Durkheim, and John Dewey during the nineteenth century, along with a host of others during the twentieth century. The term social capital itself was first used by L. T. Hanifan in 1916. In more recent years it has been popularized by, among others, Jane Jacobs, Robert Salisbury, Glen Loury, James Coleman, Robert Putnam, and Francis Fukuyama. http://en.wikipedia.org/wiki/Social_capital.

74. On this point, see especially Bowles and Gintis 2005.

75. Maynard Smith and Szathmáry 1995.

76. Hauser 2006, 2.

CHAPTER FIVE: HUMAN NATURE AND OUR BASIC NEEDS

1. Huxley 1942.

2. Dobzhansky 1962.

3. Huxley 1942, 420.

4. Williams 1966.

5. http://www.gallup.com/poll/123887/u.s.-diabetes-rate-climbs-above-11-could-hit-15–2015.aspx.

6. Harsanyi 1982, 55.

7. Heywood 1999, 298.

8. Campbell, Converse, and Rogers 1976, 9.

9. Allardt 1973, 267, 272.

10. Rist 1980, 241.

11. Baumol and Blinder 1994, 425.

12. Power in America, Wealth, Income, and Power, by G. William Domhoff, University of California, Santa Cruz, September 2005 updated October 2009. http://sociology.ucsc.edu/whorulesamerica/power/wealth.html.

13. Young 1994. Other commonly used principles include priority, consistency, competitive allocations, interpersonal comparisons, and context dependency. See also the extensive work of Steven Brams on the principle of "fair division," e.g., Brams and Taylor 1996.

14. Kahneman and Tversky 2000, Kahneman, Diener, and Schwarz 2003, Kahneman et al. 2003, 2004, Dolan and Kahneman 2008.

15. Frey and Stutzer 2002.

16. Maslow 1962. Another happiness researcher, economist Richard Layard (2005), claims that all the many kinds of satisfactions and dissatisfactions in life, like Jeremy Bentham's pains and pleasures, can be reduced to happiness. Layard, like his utilitarian forebear, calls the greatest happiness for the greatest number the "common good." One serious problem with using self-reports from happiness surveys as an objective measure of well-being is that our subjective experience is affected by so many situational factors, from genetically based personality differences to how well off we are relative to our neighbors and the status of our love lives. In my judgment, happiness provides an unreliable indicator of objective needs satisfaction.

17. See especially Sen 1982, 1985, 1992, 1993.

18. Sen 1982, 99.

19. Quoted in Nussbaum 1988, 152.

20. Sen 1993, 41. See also Dréze, Sen, and Hussain 1995.

21. Scanlon 1993.

22. Ogburn 1929, Bauer 1966.

23. Olson 1969, 97.

24. See, for example, Mazess 1978, Geist 1978, McHale and McHale 1978, Hicks and Streeten 1979.

25. Doyal and Gough 1991, 154.

26. Nussbaum and Sen 1993, 4.

27. Doyal and Gough 1991, 154.

28. Quoted in Corning 1969; also, Encyclopaedia Britannica Profiles, http://www.britannica.com/presidents/article-9116954.

29. Corning 1983.

30. Corning 2005.

31. Several other preliminary points should be mentioned briefly. First, it is important to stress the difference between primary needs and instrumental needs. As the term implies, instrumental needs serve our primary needs. Some instrumental needs are so pervasive as to be close to primary needs in their importance: for example, energy sources, protective shelter, basic utensils and tools, clothing, language skills, and so forth. Such instrumental needs are in fact the central concern of many recent efforts to develop basic needs indicators. In my view, many of the items in these lists are not primary needs at all but actually refer to instrumental needs. This is intended not to slight instrumental needs or diminish their importance but to categorize them properly with respect to their functional significance.

It should also be emphasized that instrumental needs can vary widely depending on the context. For instance, the instrumental need for a means of waste removal can range from dug latrines to open sewers and the latest high-technology waste treatment plants. Likewise, telephones and automobiles may be of little use in a simple folk society but may constitute an instrumental need in the strict sense for people who live in a complex developed society. Also, some instrumental needs take the form of economic goods and services while others relate to features of the cultural environment like law and order.

It is also important to recognize that, over the long sweep of our cultural and technological evolution, our various primary needs have generated complex hierarchies of instrumental needs. Our need for mobility, for instance, has resulted not only in the invention of automobiles but in the creation of additional instrumental needs for auto mechanics, paved roads, stop signs, the oil industry, gas stations, highway patrols, traffic courts, and so on. In fact, many inventions have been catalysts for others. If necessity is the mother of invention, as the old saying goes, the reverse is also true: inventions are the mother of necessity. Moreover, many of our instrumental technologies involve complex networks of economic interdependency. Take away the tire industry, for example, and it would cripple the automobile and trucking industries and very likely devastate our economy.

Likewise, d*ependencies* are induced, often needs unrelated to survival, some of which may even be destructive (such as an addiction to heroin, alcohol, or sugar; or compulsive gambling; or smoking). *Perceived needs* are those "preferences" that individuals think they need, regardless of the actual situation. And *wants* reflect the individual's less urgent motivations, goals, and aspirations, very possibly unrelated to any biological need. Of course, these categories often overlap. For example, a person's primary nutritional needs for protein, carbohydrates, and various vitamins and minerals may lead to the selection of a particular instrumental object (say a Big Mac), which could also become a dependency if one developed a strong psychological craving for Big Macs, or if Big Macs were the only food available. The person might also accurately perceive Big Macs as an instrumental means and, what's more, might actually enjoy them—adding to a sense of well-being.

It's also important to stress that primary needs vary, but not as much as the relativists imply. Nor are the variations a result of personal whim. The obvious case in point is nutritional needs, which are known to vary systematically (and to a substantial degree predictably) as a function of genetic and physical endowment, age, sex, reproductive status, and level of physical activity. Indeed, our nutritional needs vary not just in the number of calories but also in a broad range of required nutrients. Nevertheless, adequate nutrition constitutes a universal primary need.

Second, a complex set of interrelationships exists among the various primary needs: all needs are not equally urgent at all times, and there is an implicit hierarchy. This greatly affects the organization of our behavioral systems and the patterning of our daily activity cycles. For example, if a person's life or physical safety was suddenly threatened during a meal, one can confidently predict that he or she would stop eating. Likewise, we routinely, and at times even mindlessly, interrupt other activities to respond to the promptings of hunger, thirst, fatigue, discomfort or pain, a physical threat, the need for waste elimination, and the like.

Third, there are many interactions among our primary needs. Even though they cannot be reduced to one another, neither are they entirely independent. For example, communications (information flows) are at once an irreducible primary need and a prerequisite for the satisfaction of other primary needs—nutrition, physical safety, physical health, effective nurturance of the young—not to mention facilitating gainful employment. Likewise, waste elimination is a primary need that can also have an impact on our physical health, just as a lack of proper nutrition, sleep, or satisfactory social relationships may affect one's mental health, or physical health, or both.

Fourth, there are many potential conflicts between our needs. The obvious examples are situations where physical safety or physical health might have to be jeopardized in order to obtain food or other necessities, or where one's personal nutrition, health, and safety might have to be sacrificed for the sake of one's offspring. Other things being equal, however, the individuals who are best able to satisfy the entire gamut of primary needs will be better adapted and more likely to be successful in reproducing well-adapted offspring.

Fifth, it is important to distinguish between the analytical problems associated with determining more precisely what our basic needs are (and their functional relationship to survival and reproduction) and the more "applied" problem of how best to measure needs satisfaction for the purposes of social intelligence and social indicators. Our knowledge in many basic needs domains is still far from perfect, and

I do not underestimate the problems involved in establishing more precise criteria for each need. In the end, any "objective" measure will be only as good as the state of the art in the biological, behavioral, and social sciences. Accordingly, the Survival Indicators paradigm is a work in progress, not the actualization of some Platonic ideal.

A further point is that the Survival Indicators paradigm is designed primarily to measure current adaptation. It is not explicitly future-oriented, even though it is certainly relevant to future adaptive success. The current level of basic needs satisfaction is obviously a factor. However, the fact remains that this framework is not designed to make forecasts. It cannot anticipate such contingencies as automobile accidents, lightning bolts, earthquakes, tsunamis, plagues, wars, or asteroid strikes—much less the effects of global warming. It can only enable us to make various "if-then" predictions.

Another important point is that a basic needs approach to measuring adaptation is not the same as an explanation of culture solely in terms of basic needs. There is much more to it than that. Nor does it follow that every aspect of a cultural system is adaptive. A particular item of culture may be either adaptive, neutral, or maladaptive in relation to basic needs. Some items may be more or less directly related to a particular need. Other cultural items may be only indirectly related. (How do we account for sidewalks, or umbrellas?) Still other items may be apparently unrelated. (Can anyone provide an adaptive explanation for television game shows, baseball, or amusement parks?) Indeed, leisure "pastimes" are directly survival-relevant only for people whose livelihoods depend on them.

32. See Corning 2005. Also http://www.stanford.edu/-dement/overview-ncsdr.html.

33. Freud 1961.

34. Szasz 1961.

35. Among the many recent volumes on mental illness, see especially the American Psychological Association's *Diagnostic and Statistical Manual of Mental Disorders.*

36. http://www.bls.gov/cex/.

37. http://hdr.undp.org/en/, also http://hdr.undp.org/en/statistics/indices/hpi/.

38. The final report can be found at http://www.stiglitz-sen-fitoussi.fr/.

39. Ashcroft 2000.

40. http://www.bls.gov/tus/.

CHAPTER SIX: WHY CAPITALISM AND SOCIALISM ARE UNFAIR

1. BLS Economic News Release, December 4, 2009, http://www.bls.gov/news.release/empsit.t12.htm.

2. U.S. Census: Income, Poverty, and Health Insurance Coverage in the United States: 2008. http://www.census.gov/prod/2009pubs/p60–236.pdf.

3. Bernstein, Brocht, and Spade-Aguilar 2002.

4. Corning 2003b.

5. Working Poor Families Project 2008. http://www.aecf.org/~/media/Publication Files/NationalDataBriefFINAL.pdf.

6. Stiglitz 2010, 288. Also http://stats.oecd.org/Index.aspx?DataSetCode=ANHRS and http://stats.oecd.org/Index.aspx?DatasetCode=FTPTC. The comparison may be slightly overstated, because the United States has 4 percent more full-time workers in its workforce than do the Scandinavian countries on average. Elizabeth Warren quoted from a letter to Mary Jo Mullan, Vice President, Programs, F. B. Heron Foundation, November 26, 2007.

7. Nord, Andrews, and Carlson 2009.

8. *New York Times,* November 28, 2009. http://www.fns.usda.gov/fsp/faqs.htm.

9. CIA World Factbook. https://www.cia.gov/library/publications/the-world-factbook/rankorder/2091rank.html.

10. CIA World Factbook. https://www.cia.gov/library/publications/the-world-factbook/rankorder/2102rank.html.

11. Elise Gould, "Snapshot for July 16, 2008. Growing Disparities in Life Expectancy," Economic Policy Institute. http://www.epi.org/economic_snapshots/entry/webfeatures_snapshots_20080716/.

12. Ehrenreich 2001.

13. http://www.economist.com/daily/chartgallery/displayStory.cfm?story_id=E1_TNQVQTVN. Also see the related report by the OECD, http://www.oecd.org/document/53/0,3343,en_2649_33933_41460917_1_1_1_1,00.html.

14. "Power in America, Wealth, Income, and Power," by G. William Domhoff, UC Santa Cruz, September 2005 (updated October 2009). http://sociology.ucsc.edu/whorulesamerica/power/wealth.html. The economist Vilfredo Pareto thought the rich-poor gap represented an economic law, and he formulated the so-called 80–20 rule. Today that rule is more often associated with the relation between work effort and results, but Pareto focused on the distribution of wealth. The top 20 percent will always control 80 percent of the wealth, he claimed. Later researchers have found significant variations, but they agree that the basic idea is valid.

15. http://www.un.org/apps/news/story.asp?NewsID=20856&Cr=UN&Cr1=University.

16. Allegretto, Bernstein, and Mishel 2006. Also http://www.stateofworkingamerica.org/.

17. http://www.epi.org/economic_snapshots/entry/webfeatures_snapshots_20060621/.

18. Quoted in Corning 2003b.

19. Stiglitz 2010, 239.

20. Smith 1964, bk. 4, 2, 9.

21. Smith (like Darwin) often gets unfairly criticized when his ideas are misused. Smith's moral foundation was actually the Stoic philosophy of world citizenship, the good of the community as a whole, and the Christian teaching of the Golden Rule. In his words, we should "love our neighbour as we love ourselves" (*Theory of*

Moral Sentiments [1976], I.i.5.5). Moreover, according to Smith, virtue consists of exercising "self-command" over our baser impulses and having sympathy toward others (II.3.34). Indeed, self-command is essential to a civilized society (VI.iii.II). Moreover, Smith's justification for the invisible hand was that it would benefit society because the rich could not consume a much greater proportion of the necessities of life; their share would only be of better quality (IV.I.10). In other words, Smith was not endorsing a zero-sum game in which the rich get richer at the expense of the poor. See also Ritter 1954.

22. Mainstream economists might argue that utopian capitalism no longer reflects the orthodoxy that was once predominant. It's certainly true that modern economics has become more heterodox and dynamic, with many new theoretical nodes. (Some of these developments were discussed in chapter 4.) An illuminating overview of this work can be found in the 2004 book by David Colander and his colleagues, *The Changing Face of Economics: Conversations with Cutting Edge Economists.* Nevertheless, the orthodox neoclassical model retains a powerful hold on the discipline. Indeed, in one of the culminating "conversations" in the book, the Nobel Prize–winning economist Kenneth Arrow (a longtime leader in the discipline) insisted that "the ideas that people behave somewhat rationally and somewhat foresightedly are central to economics. . . . The idea of some type of rational action is still at the core of economics. . . . Similarly, the idea of competition remains important." Indeed, Arrow was skeptical about one alternative, complexity theory. "The problem with much of the complexity work," he said, is that "it doesn't seem to lead anywhere [theoretically]." Likewise, concerning the work in game theory and cooperation, he said, "The question is how to generalize all the noncompetitive elements in the world into a theory" (292, 293). He seems to be unfamiliar with the Synergism Hypothesis. See Corning 1983, 2005.

23. Baumol and Blinder 1994, 246.

24. Galbraith 2008, xiii.

25. Taylor 2001, 288.

26. Galbraith 2008, 46.

27. McCulloch 2002.

28. Smith 1964, I, 46.

29. Gowdy 1998, xvi–xvii.

30. Samuelson and Nordhaus 1992, 13. Also see the interview in Colander, Holt, and Rosser 2004, 311.

31. *New York Times,* September 1, 2009. http://www.nytimes.com/2009/09/02/us/02wage.html?_r=1.

32. Bowles 2004, 208.

33. Galbraith 2008, xi–xiii. Also Stiglitz 2010, 199.

34. Justice John Paul Stevens, in a passionate dissent that was joined by the other three justices who opposed the decision, wrote, "The only relevant thing that has changed is the composition of the court. . . . The court's decision is at war with the

views of generations of Americans. . . . While American democracy is imperfect, few outside of the majority of the court would have thought its flaws included a dearth of corporate money in politics." Quoted in the *New York Times,* January 26, 2010.

35. The film *Food, Inc.,* bookends a historical cycle that began with *The Jungle,* Upton Sinclair's famous muckraking book about the meatpacking industry, published in 1906. As a result of Sinclair's shocking revelations (in considerable measure), a political movement arose to impose tough food regulations, and by the 1950s the meatpacking industry was a model of both food safety and enlightened treatment of workers. However, in what could be considered a classic example of historical amnesia, all of this went into reverse with the rise of the fast-food industry and the growing pressure to expand food production. Today, unfortunately, the industry is back to the "jungle."

36. Bowles, Gintis, and Groves 2005, 1.

37. Bowles 2004, 123.

38. Target Training International 1990.

39. Grove 1999.

40. Quoted in http://www.Guardian.co.uk, Wednesday, October 21, 2009.

41. Samuelson and Nordhaus 1992, 43.

42. Baumol and Blinder 1994, 426.

43. Stiglitz 2002, xv.

44. Stiglitz 2002, 20.

45. Elsewhere Krugman explained that the profession was theoretically blinded to "the very possibility of catastrophic failures in a market economy. . . . Financial economists came to believe that markets are inherently stable. . . . There was nothing in the prevailing models suggesting the possibility of the kind of collapse that happened last year. . . . As I see it, the economics profession went astray because economists, as a group, mistook beauty, clad in impressive-looking mathematics, for truth." http://www.nytimes.com/2009/09/06/magazine/06Economic-t.html?pagewanted=all.

46. Available through Professor Geoffrey Hodgson: g.g.hodgson@herts.ac.uk.

47. Justin Fox in his 2009 book *The Myth of the Rational Market,* portrays economists as cultlike in their adoption of egregious assumptions that produce elegant but simplistic mathematical models, especially in relation to what he characterizes as the "sacred" efficient-markets hypothesis. The root of this malady can perhaps be traced back to Nobel laureate Milton Friedman, who in the 1950s championed the concept of "positive economics." According to Friedman, methodological rigor is the paramount objective, and economists should favor "parsimonious modeling," even to the point of forsaking realistic assumptions if the models are predictive. It is sufficient, in other words, to assume that people will seek at all times to maximize their "utilities." Economic theorist Eric Beinhocker, in *The Origin of Wealth,* also blames what he calls the "basic paradigm" of economics. "For the past century, economists have fundamentally misclassified the economy. The mainstream . . . has portrayed the economy as a system that moves from equilibrium point to equilibrium point." It's a concept borrowed from nineteenth-century physics, Beinhocker

notes. He argues that economies are in reality dynamic, evolving adaptive systems that exhibit both stasis and change (punctuated equilibriums) and have emergent collective properties. This model requires very different assumptions about how an economy works.

48. Quoted in Colander, Holt, and Rosser 2004, 92–93.

49. "Greenspan Concedes Error on Regulation," *New York Times*, October 23, 2008. Greenspan also admitted that his ideology had blinded him. "The whole intellectual edifice collapsed in the summer of last year" (quoted in Fox 2009, xii).

50. Reinhart and Rogoff 2009, 291–92.

51. Wolff 2000. Also Piketty and Saez 2003.

52. Pike 1966, 6–11. Over time, various reform acts passed the English Parliament, but progress was very slow. It was not until 1833 that children under nine were banned from the factories and ten- to thirteen-year-olds limited to forty-eight hours a week. Only in 1874 were employers finally limited to a fifty-six-hour workweek!

53. Engels 1958, 71.

54. For Simon, see http://www.cepa.newschool.edu/het/profiles/saintsimon.html. For Proudhon, see Hoffman 1972 and http://www.en.wikipedia.org/wiki/Property_is_theft!

55. For Robert Owen, see http://www.historyguide.org/intellect/owen.html and http://www.en.wikipedia.org/wiki/Robert_Owen.

56. Marx 1993.

57. Marx was not the originator of the idea, though he fleshed it out and built a case for it in various writings, especially in his notebooks, later published as a "manuscript" called "Theories of Surplus Value" and, of course, in his great work *Capital* (1993). In addition to the counterarguments I provide below, economist Lester Thurow, in *The Concise Encyclopedia of Economics*, provides five specific justifications for profits: (1) capitalists should be rewarded for delaying personal gratification to invest in an enterprise; (2) they should also be rewarded for the risks that they take; (3) there should be rewards for enterprise, entrepreneurial energy and organizing abilities; (4) some profits may be due to economic "rents," when a firm holding a monopoly can charge more than a competitive price and obtain an extra surplus; and (5) some profits may be due to market "imperfections"—when, for example, speculation drives up the price of crude oil and oil companies reap extra profits. http://www.econlib.org/library/CEE.html.

58. See especially Pease 1925, Beer 1940, Cole 1943, and Melitz 1959.

59. Early on, the Webbs and the Fabians were cool toward the labor union movement and even opposed it because, they maintained, it was unnecessarily adversarial and a distraction from their grandiose legislative reform efforts. But eventually they made common cause with the unions and became an influential voice in the nascent Labour Party, a relationship that endures to this day. See especially Webb and Webb 1920. Also Pease 1925 and MacKenzie 1977.

60. http://www.oecd.org/document/4/0,3343,en_2649_34533_41407428_1_1_1_1,00.html.

61. https://www.cia.gov/library/publications/the-world-factbook/rankorder/2091rank.html; also https://www.cia.gov/library/publications/the-world-factbook/rankorder/2102rank.html.

62. https://www.cia.gov/library/publications/the-world-factbook/fields/2102.html; also https://www.cia.gov/library/publications/the-world-factbook/rank order/2091rank.html; also https://www.cia.gov/library/publications/the-world-factbook/fields/2172.html.

63. http://www.forbes.com/global/2001/0319/034.html.

64. Smith 1964, I, chap. 1.

65. Brown 1991.

66. http://www.statemaster.com/encyclopedia/Grover Norquist.

67. There were, for instance, the Taoists of the fifth century BC, the Essenes of the first century AD, the Adamites of twelfth-century Bohemia, and the Levellers of seventeenth-century England. In eighteenth- and nineteenth-century America, utopian communities sprouted and died in profusion. Once-visible landmarks, such as Ephrata, Plockhoy's Commonwealth, the Society of Woman in the Wilderness, the Memnonia Institute, Shalam, and the Straight-Edgers are all long forgotten. Even such longer-lived, more economically grounded communes, such as the Oneida Community and the Shakers, are now defunct. And so are the many communes that flourished for a time in the later 1960s and early 1970s, during the Vietnam War. On the other hand, the Amish, the Hutterian Brethren, and the Israeli kibbutzim provide examples of how businesslike, efficient communal societies can succeed in certain special situations.

68. Marx's prescription was a bit more complicated than is generally appreciated. In his *Critique of the Gotha Programme* (1970), Marx identified two stages in the process of realizing a communist society. In the first stage he called for strict reciprocity. Everyone would be compelled to do his or her share, and Marx seemed to make no allowance for those who were unable to contribute for whatever reason. Later on, when the ideals of socialism had been properly instilled and everyone (presumably) would voluntarily contribute as much as possible to the common cause, Marx was willing to decouple needs from contributions.

69. Gabor 2000, 328.

70. Matthews 2006.

CHAPTER SEVEN: TOWARD A BIOSOCIAL CONTRACT

1. Quoted in Esping-Andersen 1990, 33 n. 2.

2. Polanyi 2001.

3. Lasswell 1936.

4. Esping-Andersen 1990, 9.

5. Letter to Mary Gladstone, 1881. http://www.quotationspage.com/quote/33381.html.

6. Esping-Andersen 1990, 4.

7. Quoted in http://en-wikipedia.org/wiki/Third_Way_(centrism). Also, see Gid-

dens 1999. There is another Third Way that is associated most closely with the prolific writings of the American sociologist Amitai Etzioni (1993, 2000). In essence, Etzioni advances an overarching ethical agenda. His fundamental premise is that our deepest aspiration is a "good society"—one in which people are treated as "ends, not means" (echoing Kant). Markets certainly play a role in this enterprise, but a good society requires a "balance" between the marketplace, the "state," and "communities." Indeed, Etzioni argues that it is our relationships with others—loved ones, families, communities—that give life "meaning" and purpose. Etzioni does acknowledge the practical need to provide for "a rich basic minimum for all," as he puts it. But he views this only as a means for achieving larger social and ethical ends. Etzioni has also been criticized for being one-sided and overly optimistic about human nature, among other things. See also http://en.wikipedia.org/wiki/Third_Way_(centrism).

8. Smith 1964, bk. 1, 10–11, 12.

9. Hume 1978.

10. Rawls 1999.

11. To help clarify this issue, Binmore introduces the idea of a "social index"—what he calls a "worthiness coefficient"—as a tool for weighing differences in need, power, status, and effort in deciding what is fair in any given social relationship. A real-world social contract must take all these factors into account, Binmore says. He also characterizes a social index as "a device for balancing power." But this can be problematic. Very often, the root of the problem is an *imbalance* of power. The challenge, then, may be how to change the balance of power (the social index) to achieve a fair outcome.

12. Binmore 2005, 171.

13. Binmore 2005, 183.

14. Burke 1999, 368.

15. Quoted in Thivet 2008.

16. Bowles and Gintis 1998. Two of the surveys were cited in the Bowles and Gintis article. The Gallup poll was cited in Fong, Bowles, and Gintis 2005, 278.

17. Frohlich and Oppenheimer 1992, 59, 169. For replications, see Jackson and Hill 1995, Saijo, Takahashi, and Turnbull 1996, Oleson 1997, and Cruz-Doña and Martina 2000. For a recent discussion by the authors of their work, see Oppenheimer and Frohlich 2007. See also the survey in Page and Jacobs 2009.

18. http://www.achievement.org/autodoc/page/omi0bio-1, also http://en.wikipedia.org/wiki/Pierre_Omidyar.

19. The term equity theory actually refers to two distinct, though parallel, research traditions and methodologies. The older tradition is rooted in social psychology and management. It was originally focused on workplace and management issues, though its horizon has expanded over time. The newer tradition is associated with game theory in economics and political science. It is more narrowly focused on distributive justice and how to derive equitable solutions to well-defined real-world problems, like how to divide a metaphorical birthday cake. Both traditions have developed a number of useful guidelines, principles, and model solutions for some of

the classic fairness issues. On the older tradition in equity theory, see especially Walster, Walster, and Berscheid 1978, Deutsch 1985, Miller 1999, and Spector 2008. For the work in game theory, see especially Young 1994 and Brams and Taylor 1996.

20. Rawls 1999, 96.

21. Kaplow and Shavell 2002, 5. In part, Kaplow and Shavell's purpose was to mount an argument based on the principles of welfare economics against Richard Posner's 1981 claims that judicial decisions should be based on the criteria of "wealth maximization" and "efficiency."

22. Kaplow and Shavell 2002, 27.

CHAPTER EIGHT: THE FUTURE OF FAIRNESS: THE FAIR SOCIETY

1. This old truth has also been affirmed in some of the most surprising quarters. One surprise is that gloomy proponent of monarchy, Thomas Hobbes, as quoted in the chapter epigraph. Another is Charles Darwin. In *The Voyage of the "Beagle,"* his memoir about his legendary "field trip" as a young man, he mused that "if the misery of our poor be caused not by the laws of nature, but by our institutions, great is our sin" (1972, 433). Perhaps most surprising is that great champion of property rights, John Locke. Private property, he wrote, is justified only insofar as "enough and as good is left in common for others" (quoted in Binmore 2005). Recall also Adam Smith's adherence to the Golden Rule and his admonition in *The Theory of Moral Sentiments* that we must not "injure others," and the libertarian philosopher Robert Nozick's warning to us to avoid doing "harm" (his word).

2. Organisation for Economic Co-operation and Development (OECD) statistics for 2007: http://www.oecd.org/dataoecd/48/56/41494435.pdf.

3. Judson 2009.

4. Job projections are from the *New York Times,* July 23, 2009. Norway, Sweden, and Switzerland, for example, have been able to maintain unemployment levels for extended periods at 2–3 percent or below, though everyone is struggling with the current recession. One key to the "European model" is a social contract between management and labor where the unions constrain their wage demands in the interest of price stability while management invests in job growth. See Esping-Andersen 1990.

5. http://www.investopedia.com/terms/e/employmentactof1946.asp. Conservative economists who are still in thrall to Milton Friedman's "natural rate" argument against full employment will no doubt be appalled by this proposal. But to reiterate what Galbraith points out in *The Predator State,* Friedman's key assumption is unfounded. These days, the likelihood of a wage-driven inflationary spiral is vanishingly small. It should also be noted that the original full employment bill in Congress was based on the Keynesian assumption that insufficient aggregate demand was the underlying cause of unemployment and that government measures to create full employment could correct any employment shortfall. However, the final act was watered down, and the government was only encouraged to "promote" full

employment. The act also established the President's Council of Economic Advisers and the Congressional Joint Economic Committee. A later amendment to the act, the Humphrey-Hawkins Full Employment Act of 1978, restored the language in the original bill that characterized employment as an economic "right" and called on the government to pursue a full-employment policy, though no specific program was mandated and there were no sanctions included for a failure to achieve this goal.

6. Pigeon and Clark 2003, Widerquist 2005. Actually, the idea of a guaranteed income traces back to the American revolutionary Thomas Paine in 1796.

7. The Earned Income Tax Credit (EITC, or sometimes EIC) refunds a scaled portion of income tax withholding to low-income workers. It has strong bipartisan support in this country and is used in several other countries as well. Supposedly it offers an additional incentive for work, and it favors families with children (up to three). However, there are several problems with it: it does not put any additional income in the worker's pockets beyond what was already earned; it amounts to a form of forced saving without the worker's getting any interest; the worker must apply for the tax credit, and an estimated 25 percent never do and thus lose out; workers without children get very little of their withholding back; and these lump-sum payments have become a major target for loan sharks, who seduce recipients with high-interest "advances." Why not simply raise the income tax threshold for low-income workers instead of taking it away and giving it back? The workers will still have plenty of incentive to work! See en.wikipedia.org/wiki/Earned_Income_Tax_Credit.

8. http://professionals.collegeboard.com/profdownload/trends-in-student-aid 2008.pdf. President Obama's 2009 proposals to cap student loan repayments at 10 percent of income and to forgive the remainder after twenty years (ten years for those who go into public service) is only a start in addressing this problem. Equally maladaptive for the long run is what can accurately be called our personal debt crisis. As a society, we have been living beyond our means for a generation. Consider this ominous statistical comparison. In 1982 the total household debt in this country amounted to a little over 60 percent of our GDP, while our personal savings rate was about 12 percent. By 2006, household debt had climbed inexorably to 139 percent of GDP, according to the OECD, while the savings rate had fallen virtually to zero. Our total indebtedness at the end of 2008 was close to $1 trillion. Instead of saving for our future as a nation, we have been borrowing from it, and we have heavily mortgaged ourselves and our children. http://www.oecd.org/document/51/0,3343,en33873108_33873886_38626675_1_1_1_1,00.html, also http://www.creditcards.com/credit-card-news/credit-card-industry-facts-personal-debt-statistics-1276.php#debt.

9. Matthews 2006. http://www.cahi.org/cahi_contents/resources/pdf/CAHI_Medicare_Admin_Final_hPublication.pdf. Consider also the issue of mass transit. Mobility is one of our fourteen basic needs domains and is essential to the satisfaction of many of our other basic needs. In our society, mobility often requires the use of one or more costly transportation technologies, because of the distances we must travel to meet our various needs, from employment to grocery shopping and health

care. At the outset of the twentieth century, the United States had the finest passenger railroad and urban streetcar systems in the world. We led the way in low-cost, convenient mass transportation services. And in fact many of these systems were privately owned and operated. By the 1960s, thanks in part to heavy government subsidies for the airlines and the automobile industry during the intervening years, many of these low-cost mass transit systems had been dismantled. Streetcar tracks were torn up all over the country, and passenger rail service was reduced mainly to urban commuter lines. (Amtrak, the subsidized intercity hybrid rail service, has limped along for many years on a starvation budget.)

True, there have been some new capital investments in mass transit systems over the past few decades—such as the Washington subway system, the Bay Area Rapid Transit (BART) system in San Francisco, "light rail" (streetcar) lines in various smaller cities, and of course many urban bus systems. And in the 2009 economic stimulus package, funding for mass transit projects was doubled, to $8 billion. http://www.apta.com/gap/legupdatealert/2009/Pages/2009february18.aspx.

However, the backlog of projects on the drawing boards totals more than $100 billion, and the accumulation of deferred maintenance in our existing mass transit systems (and other infrastructure) has been estimated at over $2 trillion, as I noted earlier. The bulk of our capital spending on public transportation in this country is still directed toward building airports and highways for the middle class, not mass transit for the urban class and the underclass (the estimated 50 percent of our population that either does not need or cannot afford the high cost of airfares and private automobiles). See http://www.asce.org/reportcard/2005/actionplan07.cfm and http://www.asce-sf.org/index.php?option=com_content&task=view&id=505&Itemid=42.

10. http://en-wikipedia.org/wiki/Oportunidades. The program is now being emulated in some thirty other countries. The unmet challenge for Mexico and other developing nations alike is how to create enough new jobs for these better-prepared young people.

11. See the article online titled "'Whose Company Is It?' New Insights into the Debate over Shareholders vs. Stakeholders," http://knowledge.wharton.upenn.edu/article.cfm?articleid=1826).

12. Kelly, Kelly, and Gamble 1997. Some lackluster examples in recent years have cast a shadow over the idea in some quarters, but it is much too soon to write off a good idea that may have been poorly executed. The basic principle is sound, I believe. In the cases where stakeholder capitalism has been done well, both the stakeholders and the companies have benefited.

13. *Economist,* June 6, 2009, 66–68.

14. Based on a personal interview with the public relations director for Organic Valley on June 28, 2009, along with annual reports and other background materials provided by the company, supplemented by our family's personal eating habits and brand choices over many years.

15. On social entrepreneurship, see Brinckerhoff 2000. On community governance, see Bowles and Gintis 2005. Social entrepreneurship has been defined as the

use of entrepreneurial principles and techniques to address social problems, with success measured in terms of results as well as economic viability. Bowles and Gintis define community governance as activities and enterprises that rely on "social capital" (trust and reciprocity) and that are cooperatively organized and managed. They cite as models various Chicago neighborhood associations, Japanese fishing cooperatives, and worker-owned plywood firms on the West Coast of this country. Some of the inspiration for the social capital "movement" comes from political scientist Robert Putnam's important 2000 book *Bowling Alone: The Collapse and Revival of American Community.*

16. Like so many other institutions and industries in our society, the building and loan and savings and loan banks were deregulated in the 1980s, with ultimately disastrous results. Deregulation opened the door to various fraudulent schemes, an epidemic of failures, and a financial crisis that contributed to the 1990–91 economic recession. These small local institutions have never really recovered their once-important role in the housing market.

17. http://www.muhammadyunus.org/. Also www.grameen-info.org/index.php?option=com_content.

18. http://www.cwcid.org/aboutus.htm.

19. http://www.ClintonGlobalInitiative.org. The CGI strategy is to enlist support from the full spectrum of institutions, from social entrepreneurs to NGOs, private investors, banks, and governments.

20. Akerlof and Shiller 2009, ix–x.

21. Miller 2003, xi.

22. Singer 1981. As Singer put it, "The sphere of altruism has broadened from the family and tribe to the nation, race, and now all human beings" (170). See also de Waal 2009 and Rifkin 2010.

23. Many economists hold that "scarcity" is the fundamental challenge of economics as a science. How do we use our limited resources (including especially our labor) as productively—efficiently—as possible, in order to maximize the benefits for ourselves and society? How do we divide up and distribute the goods and services we produce, which are ultimately limited in quantity? What could be called the "standard definition" of economics (it has graced the classic Paul Samuelson and William Nordhaus textbook for over fifty years) is, "Economics is the study of how societies use scarce resources to produce valuable commodities and distribute them among different people."

In an Eden of affluence, Samuelson and Nordhaus tell us, all goods would be free, although this is not strictly speaking true even in theory. At the minimum, time and energy are required to extract, transport, process, and consume any material resource or provide any service. And there are always strict limits on how much time we have available for these activities. Conversely, some resources are virtually unlimited in quantity—solar energy and seawater, for example. The problem is how to extract and use them "efficiently" to satisfy our needs. In any case, it is clear that unlimited material abundance is just another utopian fantasy.

Where the issue of fairness enters into the discussion has to do with the claim made by free market enthusiasts that markets are the most effective system for dealing with scarcity and that the price "mechanism" works efficiently to balance the relation between "supply" and "demand." There is obviously much truth to this assertion, certainly on the production side of the equation. But the additional argument that markets distribute the "goods" fairly—according to merit—is true only if you ignore our basic needs and, equally important, deny the fact that economic power pervades and often distorts the marketplace. Recall Bill Gates's comment that markets don't work for people who don't have money.

Only if the claims for merit, whether legitimate or not, are allowed to trump the claims for our basic needs can the market mechanism be called fair. But this reverses the priorities I discussed earlier and contradicts the "prime directive." Under the terms of a biosocial contract, a society must first ensure that the basic needs of its citizens are met, either by providing sufficient "purchasing power" (as the economists like to call it) or by supplying the needed goods and services directly. We know this is possible, since we are just about the only wealthy society that does not fully honor this commitment.

An economist might object that this directive ignores conditions of absolute scarcity. In fact, scarcities of various kinds happen frequently enough in any complex economy, sometimes traumatically so, and market prices respond accordingly. Conservative economists defend market dynamics as an appropriate way to ration scarcity. After all, the market is impartial, or so they claim. Those who can afford gasoline at five dollars a gallon or more will pay whatever it costs, and the rest of us are out of luck. However, this is not luck as I characterized it in chapter 2. Moreover, if we are talking about our basic needs, not just tickets to the Super Bowl, then the market mechanism may become the cause of more or less serious, even fatal "harm." The poor may starve or go without shelter against the winter cold.

One traditional alternative to allowing markets to allocate scarcity is for governments to impose some form of rationing that is deemed to be more "equitable" to everyone. To be sure, attempts to impose a political solution are frequently undermined by cheating—"black markets" that subvert the basic goal by surreptitiously reimposing a market pricing mechanism. But this outcome is not inevitable. During World War II, for example, a wide variety of consumer items in this country were strictly and successfully rationed, from gasoline to butter, meat, and more. Wartime patriotism and an attitude of "we're all in this together" were key factors in the overall success of the rationing program, along with competent management and aggressive enforcement when black market sales did occur. In other words, rationing can work if there is a strong political consensus and a collective will to make it work.

The other alternative, more commonly used these days, is to provide public subsidies for the poor, such as heating oil during the winter. One problem with this is that the near-poor who don't qualify for subsidies and don't have a lot of discretionary income may suffer disproportionately. The other problem is that the taxpayers must pick up the tab, with the burden, again, falling disproportionately on the middle class.

24. Savage 2001, 19.

25. Quoted in Buss 2005, 768.

26. Wilkinson and Pickett 2009. For example, America has the lowest life expectancy, by far the highest crime rates, and two and one-half times the rate of mental illness of any other developed nation. This is not surprising. We also have the biggest gap between the rich and the poor and the highest poverty rate.

27. Quoted in *The Progress Report,* March 16, 2010, http://www.americanprogressaction.org. There was, of course, much pushback against this warped view. For example, a newspaper cartoon captioned, "If Jesus Returned as Glenn Beck," showed Jesus in three panels preaching: "Don't heal the sick, that's socialism," "Don't shelter the homeless, that's communism," and "Don't feed the hungry, that's Nazism."

28. Yogi Berra is famous for many quirky sayings, such as, "If you don't know where you're going, you might end up someplace else," "It's like déjà vu all over again," "Always go to other people's funerals, otherwise they won't come to yours," "You can observe a lot just by watching," "It ain't over till it's over," "Nobody goes there anymore, it's too crowded," "It's tough to make predictions, especially about the future," and "If the world was perfect, it wouldn't be." But as Berra himself admitted, "I didn't say everything I said."

References

Akerlof, George A. 2007. The Missing Motivation in Macroeconomics. *American Economic Review* 97 (1): 5–36.

Akerlof, George A., and Robert J. Shiller. 2009. *Animal Spirits: How Human Psychology Drives the Economy, and Why It Matters for Global Capitalism.* Princeton, NJ: Princeton University Press.

Alejandro, Roberto D. 1998. *The Limits of Rawlsian Justice.* Baltimore, MD: Johns Hopkins University Press.

Alexander, Richard D. 1987. *The Biology of Moral Systems.* New York: Aldine de Gruyter.

Allardt, Erik. 1973. Individual Needs, Social Structures, and Indicators of National Development. In *Building, States and Nations: Models and Data Resources,* ed. Shmuel N. Eisenstadt and Stein Rokkan, 259–76. Beverly Hills, CA: Sage.

Allegretto, Sylvia A., Jared Bernstein, and Lawrence Mishel. 2006. *The State of Working America, 2006–07.* Ithaca, NY: ILR Press.

Aristotle. 1946/350 BC. *The Politics.* Trans. Ernest Barker. Oxford: Oxford University Press.

____. 1985/350 BC. *Nichomachean Ethics.* Trans. Terence Irwin. Indianapolis: Hackett.

Arnhart, Larry. 1998. *Darwinian Natural Right: The Biological Ethics of Human Nature.* Albany: State University of New York Press, 1998.

Ashcroft, Frances. 2000. *Life at the Extremes.* Berkeley: University of California Press.

Axelrod, Robert. 1984. *The Evolution of Cooperation.* New York: Basic Books.

Axelrod, Robert, and William Hamilton. 1981. The Evolution of Cooperation. *Science* 211:1390.

Barker, Ernest. 1960/1918. *Greek Political Theory.* London: Methuen.

Bauer, Raymond A., ed. 1966. *Social Indicators.* Cambridge, MA: MIT Press.

Baumol, William J., and Alan S. Blinder. 1994. *Economics: Principles and Policy.* 6th ed. Fort Worth, TX: Dryden Press.

Bazzett, Terence J. 2008. *An Introduction to Behavior Genetics.* Sunderland: Sinauer.

Bear, Mark F., Barry W. Connors, and Michael A. Paradiso. 2007. *Neuroscience: Exploring the Brain.* 3rd ed. Philadelphia: Lippincott, Williams and Wilkins.

Beer, Max. 1940. *A History of British Socialism.* London: Allen and Unwin.

Beinhocker, Eric D. 2006. *The Origin of Wealth: Evolution, Complexity and the Radical Remaking of Economics.* Boston: Harvard Business School Press.

Bellah, Robert W. 1991. *The Good Society.* New York: Alfred A. Knopf.

____. 1995. *Community Properly Understood: A Defense of Democratic Communitarianism.* Washington, DC: Communitarian Network/George Washington University Institute for Communitarian Policy Studies.

Bentham, Jeremy. 1970/1789. *Introduction to the Principles of Morals and Legislation.* London: Athlone.

Bernstein, Jared, Chauna Brocht, and Maggie Spade-Aguilar. 2002. *How Much Is Enough? Basic Family Budgets for Working Families.* Washington, DC: Economic Policy Institute.

Binmore, Ken. 2005. *Natural Justice.* Cambridge, MA: MIT Press.

Boardman, John, Jasper Griffin, and Oswyn Murray, eds. 2002. *Oxford History of Greece and the Hellenistic World.* 2nd ed. New York: Oxford University Press.

Boehm, Christopher. 1999. *Hierarchy in the Forest: The Evolution of Egalitarian Behavior.* Cambridge, MA: Harvard University Press.

Bouchard, Thomas J. 2004. Genetic Influence on Human Psychological Traits. *Current Directions in Psychological Science* 13 (4): 148–51.

Bowles, Samuel. 2004. *Microeconomics: Behavior, Institutions, and Evolution.* New York: Russell Sage Foundation/Princeton University Press.

Bowles, Samuel, and Herbert Gintis. 1998. Is Equality Passé? Homo Reciprocans and the Future of Egalitarian Politics. *Boston Review,* Fall, 4–10.

____. 2005. Social Capital, Moral Sentiments, and Community Governance. In *Moral Sentiments and Material Interests: The Foundations of Cooperation in Economic Life,* ed. Herbert Gintis, Samuel Bowles, Robert Boyd, and Ernst Fehr, 379–98. Cambridge, MA: MIT Press.

Bowles, Samuel, Herbert Gintis, and Melissa Osborne Groves, eds. 2005. *Unequal Chances: Family Background and Economic Success.* New York: Russell Sage Foundation/Princeton University Press.

Brams, Steven J., and Alan D. Taylor. 1996. *Fair Division: From Cake-Cutting to Dispute Resolution.* Cambridge: Cambridge University Press.

Brinckerhoff, Peter C. 2000. *Social Entrepreneurship: The Art of Mission-Based Venture Development.* New York: John Wiley.

Brown, Donald E. 1991. *Human Universals.* Philadelphia: Temple University Press.

Burke, Edmund. 1999/1847. *Works of Edmund Burke.* New York: Harper.

Burns, Jennifer. 2009. *Goddess of the Market: Ayn Rand and the American Right.* New York: Oxford University Press.

Buss, David M., ed. 2005. *The Handbook of Evolutionary Psychology.* Hoboken, NJ: John Wiley.

Campbell, Angus, Philip E. Converse, and Willard L. Rogers. 1976. *The Quality of American Life: Perceptions, Evaluations, and Satisfactions.* New York: Russell Sage Foundation.

Carneiro, Robert L. 1970. A Theory of the Origin of the State. *Science* 169:733–38.

Cicero, Marcus Tullius. 2001/45 BC. *On Moral Ends.* Cambridge: Cambridge University Press.

Cohen, Ronald L., ed. 1986. *Justice: Views from the Social Sciences.* New York: Plenum.

Colander, David, Richard P. F. Holt, and J. Barkley Rosser Jr. 2004. *The Changing Face of Economics: Conversations with Cutting Edge Economists.* Ann Arbor: University of Michigan Press.

Cole, George Douglas Howard. 1943. *Fabian Socialism.* London: Allen and Unwin.

Connolly, Peter. 1998. *The Ancient City: Life in Classical Athens and Rome.* Oxford: Oxford University Press.

Corning, Peter A. 1969. The Evolution of Medicare: From Idea to Law. Research Report 29. Office of Research and Statistics, Social Security Administration, Washington, DC.

____. 1983. *The Synergism Hypothesis: A Theory of Progressive Evolution.* New York: McGraw-Hill.

____. 2003a. *Nature's Magic: Synergy in Evolution and the Fate of Humankind.* Cambridge: Cambridge University Press.

____. 2003b. The Continuing Decline of the Middle Class. *Indicators* 2 (4): 74–78.

____. 2005. *Holistic Darwinism: Synergy, Cybernetics and the Bioeconomics of Evolution.* Chicago: University of Chicago Press.

____. 2007. Synergy Goes to War: A Bioeconomic Theory of Collective Violence. *Journal of Bioeconomics* 9 (2): 109–44.

Cruz-Doña, Rena dela, and Alan Martina. 2000. Diverse Groups Agreeing on a System of Justice and Distribution: Evidence from the Philippines. *Journal of Interdisciplinary Economics* 11:35–76.

Damasio, Antonio R. 1994. *Descartes' Error: Emotion, Reason and the Human Brain.* New York: G. P. Putnam.

Darwin, Charles. 1874/1871. *The Descent of Man, and Selection in Relation to Sex.* New York: A. L. Burt.

____. 1965/1873. *The Expression of the Emotions in Man and Animals.* London: John Murray.

____. 1972/1839. *The Voyage of the "Beagle."* New York: Bantam Books.

Dawkins, Richard. 1989/1976. *The Selfish Gene.* 2nd ed. Oxford: Oxford University Press.

Delcomyn, Fred. 1998. *Foundations of Neurobiology.* New York: Freeman.

Deutsch, Morton. 1985. *Distributive Justice.* New Haven, CT: Yale University Press.

de Waal, Frans B. M. 1982. *Chimpanzee Politics: Power and Sex among Apes.* New York: Harper and Row.

____. 1996. *Good Natured: The Origin of Right and Wrong in Humans and Other Animals.* Cambridge, MA: Harvard University Press.

____. 1997. *Bonobo: The Forgotten Ape.* Berkeley: University of California Press.

____, ed. 2001. *Tree of Origin: What Primate Behavior Can Tell Us about Human Social Evolution.* Cambridge, MA: Harvard University Press.

____. 2005. *Our Inner Ape.* New York: Riverhead Books.

____. 2006. *Primates and Philosophers: How Morality Evolved.* Princeton, NJ: Princeton University Press.

____. 2009. *The Age of Empathy: Nature's Lessons for a Kinder Society.* New York: Harmony Books.

Diamond, Jared. 1997. *Guns, Germs, and Steel: The Fates of Human Societies.* New York: W. W. Norton.

Dilalla, Lisabeth F., and Irving I. Gottesman, eds. 2004. *Behavior Genetics Principles: Perspectives in Development, Personality, and Psychopathology.* Washington, DC: American Psychological Association.

Divale, William T. 1973. *Warfare in Primitive Societies: A Bibliography.* Santa Barbara, CA: ABC-Clio.

Dobzhansky, Theodosius. 1962. *Mankind Evolving: The Evolution of the Human Species.* New Haven, CT: Yale University Press.

Dolan, Paul, and Daniel Kahneman. 2008. Interpretations of Utility and Their Implications for the Valuation of Health. *Economic Journal* 118:215–34.

Doyal, Len, and Ian Gough. 1991. *A Theory of Human Need.* London: Macmillan Education.

Dréze, Jean, Amartya Sen, and Athar Hussain, eds. 1995. *The Political Economy of Hunger.* Oxford: Clarendon Press.

Dunn, Susan, ed. 2002. *Jean-Jacques Rousseau.* New Haven, CT: Yale University Press.

Durkheim, Émile. 1938/1895. *The Rules of Sociological Method.* Chicago: University of Chicago Press.

Edgerton, Robert B. 1992. *Sick Societies: Challenging the Myth of Primitive Harmony.* New York: Free Press.

Ehrenreich, Barbara. 2001. *Nickel and Dimed: On (Not) Getting By in America.* New York: Henry Holt.

Ehrlich, Paul R. 2000. *Human Natures: Genes, Cultures, and the Human Prospect.* Washington, DC.: Island Press/Shearwater Books.

Eibl-Eibesfeldt, Irenäus. 1989. *Human Ethology.* New York: Aldine de Gruyter.

Elshtain, Jean Bethke. 1995. *Democracy on Trial.* New York: Basic Books.

Elster, Jon. 1992. *Local Justice: How Institutions Allocate Scarce Goods and Necessary Burdens.* New York: Russell Sage Foundation.

____, ed. 1995. *Local Justice in America.* New York: Russell Sage Foundation.

Ember, Carol R. 1978. Myths about Hunter-Gatherers. *Ethnology* 27:239–448.

Engels, Friedrich. 1958/1845. *Conditions of the Working Class in England.* Trans. W. O. Henderson and W. H. Chaloner. Stanford, CA: Stanford University Press.

Esping-Andersen, Gøsta. 1990. *The Three Worlds of Welfare Capitalism.* Princeton. NJ: Princeton University Press.

Etzioni, Amitai. 1993. *The Spirit of Community.* New York: Crown.

____. 1998. *The New Golden Rule: Morality and Community in a Democratic Society.* New York: Basic Books.

____. 2000. *The Third Way to a Good Society.* London: Demos.

____. 2005. The Fair Society. In *Uniting America: Restoring the Vital Center to American Democracy,* ed. Norton Garfinkle and Daniel Yankelovich, 211–23. New Haven, CT: Yale University Press.

Ewald, Paul W. 1991. Transmission Modes and the Evolution of Virulence: With Special Reference to Cholera, Influenza and AIDS. *Human Nature* 2 (1): 1–30.

Fagan, Brian M. 1998. *People of the Earth: An Introduction to World Prehistory.* 9th ed. New York: Addison Wesley Longman.

Field, Alexander. 2001. *Altruistically Inclined: The Behavioral Sciences, Evolutionary Theory and the Origins of Reciprocity.* Ann Arbor: University of Michigan Press.

Flack, Jessica, and Frans B. M. de Waal. 2000. "Any Animal Whatever": Darwinian Building Blocks of Morality in Monkeys and Apes. *Journal of Consciousness Studies* 7 (1–2): 1–29.

Fong, Christina M., Samuel Bowles, and Herbert Gintis. 2005. Reciprocity and the Welfare State. In *Moral Sentiments and Material Interests: The Foundations of Cooperation in Economic Life,* ed. Herbert Gintis, Samuel Bowles, Robert Boyd, and Ernst Fehr, 277–302. Cambridge, MA: MIT Press.

Fox, Justin. 2009. *The Myth of the Rational Market: A History of Risk, Reward, and Delusion on Wall Street.* New York: HarperCollins.

Frank, Robert H. 1988. *Passions within Reason: The Strategic Role of the Emotions.* New York: W. W. Norton.

Freud, Sigmund. 1961/1930. *Civilization and Its Discontents.* In *The Standard Edition of the Complete Psychological Works of Sigmund Freud,* trans. J. Strachey, vol. 21. New York: W. W. Norton.

Frey, Bruno S., and Alois Stutzer. 2002. *Happiness and Economics: How the Economy and Institutions Affect Well-Being.* Princeton, NJ: Princeton University Press.

Frohlich, Norman, and Joe A. Oppenheimer. 1992. *Choosing Justice: An Experimental Approach to Ethical Theory.* Berkeley: University of California Press.

Gabor, Andrea. 2000. *The Capitalist Philosophers.* New York: Times Books/Random House.

Galbraith, James K. 1998. *Created Unequal: The Crisis in American Pay.* New York: Free Press.

____. 2008. *The Predator State: How Conservatives Abandoned the Free Market and Why Liberals Should Too.* New York: Free Press.

Gat, Azar. 2006. *War in Human Civilization.* New York: Oxford University Press.

Gazzaniga, Michael S. 2005. *The Ethical Brain.* New York: Dana Press.

Geist, Valerius. 1978. *Life Strategies, Human Evolution, Environmental Design: Toward a Biological Theory of Health.* New York: Springer-Verlag.

Giddens, Anthony. 1999. *The Third Way: The Renewal of Social Democracy.* Malden, MA: Polity Press.

Gintis, Herbert. 2008. Punishment and Cooperation. *Science* 319:1345–46.

Gintis, Herbert, Samuel Bowles, Robert Boyd, and Ernst Fehr, eds. 2005. *Moral Sentiments and Material Interests: The Foundations of Cooperation in Economic Life.* Cambridge, MA: MIT Press.

Gintis, Herbert, and Ernst Fehr. 2007. Human Nature and Social Cooperation. *Annual Review of Sociology* 33 (3): 1–22.

Gladwell, Malcolm. 2002. *The Tipping Point.* New York: Back Bay Books.

Gluckman, Max. 1940. The Kingdom of the Zulu of South Africa. In *African Political Systems,* ed. Meyer Fortes and Edward E. Evans-Pritchard, 25–55. London: Oxford University Press.

____. 1969. The Rise of a Zulu Empire. *Scientific American* 202 (4): 157–68.

Gouldner, Alvin W. 1960. The Norm of Reciprocity: A Preliminary Statement. *American Sociological Review* 25:161–78.

Gowdy, John M., ed. 1998. *Limited Wants, Unlimited Means: A Reader on Hunter-Gatherer Economics and the Environment.* Washington, DC: Island Press.

Grove, Andrew. 1999. *Only the Paranoid Survive.* New York: Broadway Business Books.

Haidt, Jonathan. 2007. The New Synthesis in Moral Psychology. *Science* 316:998–1002.

Harris, Edward M. 2006. *Democracy and the Rule of Law in Classical Athens: Essays on Law, Society, and Politics.* New York: Cambridge University Press.

Harsanyi, John D. 1982. Morality and the Theory of Rational Behavior. In *Utilitarianism and Beyond,* ed. Amartya K. Sen and Bernard W. Williams, 39–62. Cambridge: Cambridge University Press.

Hauser, Marc D. 2006. *Moral Minds: How Nature Designed Our Universal Sense of Right and Wrong.* New York: HarperCollins.

Henrich, Joseph, Robert Boyd, Samuel Bowles, Colin Camerer, Herbert Gintis, and Richard McElreath. 2001. In Search of Homo Economicus: Behavioral Experiments in Fifteen Small-Scale Societies. *American Economic Review* 91:73–78.

Henrich, Joseph, Jean Ensminger, Richard McElreath, Abigail Barr, Clark Barrett, Alexander Bolyantz, Juan Camilo Cardenas, Michael Gurven, Edwins Gwako, Natalie Henrich, Carolyn Lesorogol, Frank Marlowe, David Tracer, and John Ziker. 2010. Markets, Religion, Community Size, and the Evolution of Fairness and Punishment. *Science* 327 (March): 1480–84.

Heywood, Andrew. 1999. *Political Theory: An Introduction.* London: Macmillan.

Hicks, Norman, and Paul Streeten. 1979. Indicators of Development: The Search for a Basic Needs Yardstick. *World Development* 7:567–80.

Hobbes, Thomas. 1928/1640. *The Elements of Law, Natural and Politic.* New York: Macmillan.

____. 1962/1651. *Leviathan, or The Matter, Form, and Power of a Commonwealth Ecclesiastical and Civil.* New York: Collier Books.

Hoffer, Eric. 1951. *The True Believer.* New York: Harper and Row.

Hoffman, Robert L. 1972. *Revolutionary Justice: The Social and Political Theory of Pierre-Joseph Proudhon.* Urbana: University of Illinois Press.

Hubel, David H., and Torsten N. Wiesel. 1979. Brain Mechanisms of Vision. *Scientific American* 241:130–44.

Huff, Darrell. 1954. *How to Lie with Statistics.* New York: W. W. Norton.

Hume, David. 1978/1739. *A Treatise of Human Nature.* 2nd ed. Ed. L. A. Selby-Bigge. Oxford: Clarendon Press.

Huxley, Julian S. 1942. *Evolution: The Modern Synthesis.* New York: Harper and Row.

Huxley, Thomas Henry. 2005/1893. *Collected Essays,* vol. 1. Whitefish, MT: Kessinger.

Jackson, Michael, and Peter Hill. 1995. A Fair Share. *Journal of Theoretical Politics* 7:69–79.

Joyce, Richard. 2007. *The Evolution of Morality.* Cambridge, MA: MIT Press, Bradford Books.

Judson, Bruce. 2009. *It Could Happen Here: America on the Brink.* New York: HarperCollins.

Kahneman, Daniel, Ed Diener, and Norbert Schwarz, eds. 2003. *Well-Being: The Foundations of Hedonic Psychology.* New York: Russell Sage Foundation.

Kahneman, Daniel, Alan B. Krueger, David Schkade, Norbert Schwarz, and Arthur Stone. 2004. Toward National Well-Being Accounts. *American Economic Review* 94:429–34.

Kahneman, Daniel, and Amos Tversky, eds. 2000. *Choices, Values and Frames.* New York: Cambridge University Press and Russell Sage Foundation.

Kant, Immanuel. 1997/1785. *Groundwork of the Metaphysics of Morals.* Trans. Mary Gregor. Cambridge: Cambridge University Press.

____. 1999/1788. *Critique of Pure Reason.* New York: Cambridge University Press.

Kaplow, Louis, and Steven Shavell. 2002. *Fairness versus Welfare.* Cambridge, MA: Harvard University Press.

Katz, Leonard D., ed. 2000. *Evolutionary Origins of Morality.* Thorverton, UK: Imprint Academic.

Keeley, Lawrence H. 1996. *War Before Civilization: The Myth of the Peaceful Savage.* Oxford: Oxford University Press.

Kelly, Gavin, Dominic Kelly, and Andrew Gamble, eds. 1997. *Stakeholder Capitalism.* New York: St. Martin's Press.

Keynes, John Maynard. 1936. *The General Theory of Employment, Interest and Money.* Cambridge: Cambridge University Press.

Kingdon, Jonathan. 1993. *Self-Made Man: Human Evolution from Eden to Extinction?* New York: John Wiley.

Klein, Richard G. 1999. *The Human Career: Human Biological and Cultural Origins.* 2nd ed. Chicago: University of Chicago Press.

Klosko, George. 1992. *The Principles of Fairness and Political Obligation.* Lanham, MD: Rowman and Littlefield.

Kohlberg, Lawrence. 1981. *Essays on Moral Development,* vol. 1, *The Philosophy of Moral Development.* San Francisco: Harper and Row.

Kohlberg, Lawrence, Charles Levine, and Alexandra Hewer. 1983. *Moral Stages: A Current Formulation and a Response to Critics.* Basel: Karger.

Kolm, Serge-Christophe. 1996. *Modern Theories of Justice.* Cambridge, MA: MIT Press.

Krugman, Paul. 2009. *The Return of Depression Economics and the Crisis of 2008.* New York: W. W. Norton.

Kummer, Hans. 1968. *Social Organization of Hamadryas Baboons: A Field Study.* Chicago: University of Chicago Press.

Lasswell, Harold. 1936. *Politics: Who Gets What, When, How.* New York: McGraw-Hill.

Layard, Richard. 2005. *Happiness: Lessons from a New Science.* New York: Penguin Press.

Lippmann, Walter. 1989/1954. *The Public Philosophy.* Edison, NJ: Transaction Books.

Locke, John L. 1970/1690. *Two Treatises of Government.* Ed. P. Laslett. Cambridge, MA: Harvard University Press.

MacCormack, Geoffrey. 1976. Reciprocity. *Man,* n.s., 11 (1): 89–103.

Machiavelli, Niccolò. 1950/1531. *Discourses.* New Haven, CT: Yale University Press.

____. 1996/1513. *The Prince.* Atlantic Highlands, NJ: Humanities Press.

MacIntyre, Alasdair. 1981. *After Virtue: A Study in Moral Theory.* Notre Dame, IN: University of Notre Dame Press.

____. 1988. *Whose Justice? Which Rationality?* Notre Dame, IN: University of Notre Dame Press.

MacKenzie, Norman Ian. 1977. *The Fabians.* New York: Simon and Schuster.

Macmillan, Malcolm. 2002. *An Odd Kind of Fame: Stories of Phineas Gage.* Cambridge, MA: MIT Press.

____. 2008. Phineas Gage—Unraveling the Myth. *Psychologist* 21 (9): 828–31.

Maisels, Charles Keith. 1999. *Early Civilizations of the Old World: The Formative Histories of Egypt, the Levant, Mesopotamia, India, and China.* London: Routledge.

Marx, Karl. 1970/1875. Critique of the Gotha Programme. In *Marx/Engels Selected Works,* 3:13–30. Moscow: Progress.

____. 1993/1867. *Capital.* Provo, UT: Regal.

____. 2004/1848. *Communist Manifesto.* Peterborough, ON: Broadview Press.

Maryanski, Alexandra, and Jonathan H. Turner. 1992. *The Social Cage: Human Nature and the Evolution of Society.* Stanford, CA: Stanford University Press.

Maslow, Abraham M. 1962. *Toward a Psychology of Being.* Princeton, NJ: Van Nostrand.

Masters, Roger D. 1989. *The Nature of Politics.* New Haven, CT: Yale University Press.

Masters, Roger D., and Margaret Gruter, eds. 1992. *The Sense of Justice: Biological Foundations of Law.* Newbury Park, CA: Sage.

Matthews, Merrill. 2006. *Medicare's Hidden Administrative Costs: A Comparison of Medicare and the Private Sector.* Alexandria, VA: Council for Affordable Health Insurance.

Maynard Smith, John, and Eörs Szathmáry. 1995. *The Major Transitions in Evolution.* Oxford: Freeman.

Mazess, Richard B. 1978. Adaptation: A Conceptual Framework. In *Evolutionary Models and Studies in Human Diversity,* ed. Robert J. Meier, Charlotte M. Otten, and Fathi Abdel-Hameed, 9–15. The Hague: Mouton.

McCulloch, J. R., ed. 2002. *The Works of David Ricardo.* Honolulu: University Press of the Pacific.

McCullough, David. 1993. *Truman.* New York: Simon and Schuster.

McHale, John, and Magda C. McHale. 1978. *Basic Human Needs: A Framework for Action.* New Brunswick, NJ: Transaction Books.

McShea, Robert J. 1990. *Morality and Human Nature: A New Route to Ethical Theory.* Philadelphia: Temple University Press.

Melitz, Jack. 1959. Trade Unions and Fabian Socialism. *Industrial and Labor Relations Review* 12 (4): 554–67.

Milgram, Stanley. 1974. *Obedience to Authority: An Experimental View.* New York: Harper and Row.

Miller, David. 1976. *Social Justice.* Oxford: Oxford University Press.

____. 1999. *Principles of Social Justice.* Cambridge, MA: Harvard University Press.

Miller, Matthew. 2003. *The Two Percent Solution: Fixing America's Problems in Ways Liberals and Conservatives Can Love.* New York: Perseus Books.

Moore, Barrington, Jr. 1978. *Injustice: The Social Basis of Obedience and Revolt.* New York: M. E. Sharpe.

Morris, Donald R. 1965. *The Washing of the Spears.* New York: Simon and Schuster.

Nord, Mark, Margaret Andrews, and Steven Carlson. 2009. Household Food Security in the United States, 2008. Economic Research Report ERR-83. United States Department of Agriculture.

Nowak, Martin, and Karl Sigmund. 1993. A Strategy of Win-Stay, Lose Shift That Outperforms Tit-for-Tat in the Prisoner's Dilemma Game. *Nature* 364:56–58.

Nozick, Robert. 1974. *Anarchy, State and Utopia.* Oxford: Blackwell.

Nussbaum, Martha. 1988. Nature, Function and Capability: Aristotle on Political Distribution. *Oxford Studies in Ancient Philosophy,* suppl., 145–84.

Nussbaum, Martha, and Amartya Sen, eds. 1993. *The Quality of Life.* Oxford: Clarendon Press.

Ogburn, William F., ed. 1929. *Social Changes in 1928.* Chicago: University of Chicago Press.

Oleson, Paul E. 1997. An Experimental Examination of Alternative Theories of Distributive Justice and Economic Fairness. Paper presented at the Public Choice Society Meetings, San Francisco.

Olson, Mancur. 1969. *Toward a Social Report.* Washington, DC: U.S. Government Printing Office.

Oppenheimer, Joe A., and Norman Frohlich. 2007. Demystifying Social Welfare: Needs and Social Justice in the Evaluation of Democracies. *Maryland Law Review* 67 (1): 85–122.

Page, Benjamin I., and Lawrence R. Jacobs. 2009. *Class War: What Americans Really Think about Economic Inequality*. Chicago: University of Chicago Press.

Pease, Edward. 1925. *A History of the Fabian Society.* 2nd ed. London: Allen and Unwin.

Pettit, Philip. 1980. *Judging Justice: An Introduction to Contemporary Political Philosophy.* London: Routledge and Kegan Paul.

Pfaff, Donald W. 2007. *The Neuroscience of Fair Play: Why We (Usually) Follow the Golden Rule.* New York: Dana Press.

Pfeiffer, John E. 1977. *The Emergence of Society: A Pre-history of the Establishment.* New York: McGraw-Hill.

Phelps, Edmund S. 1997. *Rewarding Work: How to Restore Participation and Self-Support to Free Enterprise.* Cambridge, MA: Harvard University Press.

Piaget, Jean. 1932. *The Moral Judgment of the Child.* London: Kegan Paul, Trench, Trubner.

Pigeon, Marc-Andre, and Charles M. A. Clark. 2003. *The Basic Income Guarantee: Ensuring Progress and Prosperity in the 21st Century.* Dublin: Liffey.

Pike, E. Royston. 1966. *"Hard Times": Human Documents of the Industrial Revolution.* New York: Praeger.

Piketty, Thomas, and Emmanuel Saez. 2003. Income Inequality in the United States, 1913–1998. *Quarterly Journal of Economics* 68 (1): 1–39.

Pinker, Steven. 2002. *The Blank Slate: The Modern Denial of Human Nature.* New York: Viking Penguin.

Plato. 1946/380 BC. *The Republic.* Trans. B. Jowett. Cleveland, OH: World.

____. 1992a/ca. 347 BC. *The Laws.* New York: Penguin Books.

____. 1992b/360 BC. *The Statesman.* Indianapolis: Hackett.

Plomin, Robert, and John C. DeFries. 1998. The Genetics of Cognitive Abilities and Disabilities. *Scientific American* 278 (5): 62–68.

Plomin, Robert, John C. Defries, and Gerald E. McClearn. 1990. *Behavioral Genetics: A Primer.* New York: W. H. Freeman.

Plutarch. 2001/ca. AD 46–120. *Lives.* New York: Modern Library.

Polanyi, Karl. 2001/1944. *The Great Transformation: The Political and Economic Origins of Our Time.* Boston: Beacon Press.

Pomeroy, Sarah B., Stanley M. Burstein, Walter Donlan, and Jennifer Tolbert Roberts. 2007. *Ancient Greece: A Political, Social and Cultural History.* 2nd ed. New York: Oxford University Press.

Posner, Richard A. 1981. *The Economics of Justice.* Cambridge, MA: Harvard University Press.

Price, T. Douglas, and Gary M. Feinman, eds. 1995. *Foundations of Social Inequality.* New York: Plenum.

Putnam, Robert D. 2000. *Bowling Alone: The Collapse and Revival of American Community.* New York: Simon and Schuster.

Rand, Ayn. 1957. *Atlas Shrugged.* New York: Random House.

____. 1993/1943. *The Fountainhead.* New York: Penguin.

Raphael, David Daiches. 2001. *Concepts of Justice.* Oxford: Clarendon Press.

Rawls, John. 1995. *Political Liberalism.* New York: Columbia University Press.

____. 1999/1971. *A Theory of Justice.* Cambridge, MA: Belknap Press.

Reich, Robert B. 2007. *Supercapitalism: The Transformation of Business, Democracy, and Everyday Life.* New York: Alfred A. Knopf.

Reinhart, Carmen M., and Kenneth S. Rogoff. 2009. *This Time Is Different: Eight Centuries of Financial Folly.* Princeton, NJ: Princeton University Press.

Ridley, Matt. 1997. *The Origins of Virtue: Human Instincts and the Evolution of Cooperation.* New York: Viking.

Rifkin, Jeremy. 2010. *The Empathic Civilization: The Race to Global Consciousness in a World in Crisis.* New York: Tarcher/Penguin.

Rist, Gilbert. 1980. Basic Questions about Basic Human Needs. In *Human Needs,* ed. Katrin Lederer, 233–53. Cambridge, MA: Oelgeschlager, Gunn and Hain.

Ritter, William Emerson. 1954. *Charles Darwin and the Golden Rule.* Ed. Edna Watson Bailey. New York: Storm.

Rousseau, Jean-Jacques. 1984/1762. *Of the Social Contract.* Trans. Charles M. Sherover. New York: Harper and Row.

____. 2004/1755. *Discourse on Inequality.* Whitefish, MT: Kessinger.

Rubin, Paul. 2002. *Darwinian Politics: The Evolutionary Origin of Freedom.* New Brunswick, NJ: Rutgers University Press.

Rushton, J. Philippe, David W. Fulker, Michael C. Neal, David K. B. Nias, and Hans J. Eysenck. 1986. Altruism and Aggression: The Heritability of Individual Differences. *Journal of Personality and Social Psychology* 50:1192–98.

Sabine, George H. 1961. *A History of Political Theory.* 3rd ed. New York: Holt, Rinehart and Winston.

Saijo, Tatsuyoshi, Shusuke Takahashi, and Stephen Turnbull. 1996. Justice in Income Distribution: An Experimental Approach. Paper presented at 1996 ISA conference, San Diego, CA.

Salter, Frank. 2008. Westermarck's Altruism: Charity Releasers, Moral Emotions and the Welfare Ethic. *Politics and the Life Sciences* 27 (2): 28–46.

Samuelson, Paul A., and William D. Nordhaus. 1992. *Economics.* 14th ed. New York: McGraw-Hill.

Sanfey, Alan G. 2007. Social Decision-Making: Insights from Game Theory and Neuroscience. *Science* 318:598–602.

Savage, Michael. 2002. *The Savage Nation: Saving America from the Liberal Assault on Our Borders, Language and Culture.* Torrance, CA: WND Books.

Scanlon, Thomas M. 1993. Value, Desire and Quality of Life. In *The Quality of Life,* ed. Martha Nussbaum and Amartya Sen, 185–205. Oxford: Clarendon Press.

Schumpeter, Joseph A. 1934. *The Theory of Economic Development.* Cambridge, MA: Harvard University Press.

Sen, Amarta K. 1982. *Choice, Welfare and Measurement.* Cambridge, : MIT Press.

____. 1985. *Commodities and Capabilities.* Amsterdam: Elsevier.

____. 1992. *Inequality Reexamined.* Oxford: Clarendon Press.

____. 1993. Capability and Well-Being. In *The Quality of Life,* ed. Martha Nussbaum and Amartya Sen, 30–53. Oxford: Clarendon Press.

Shultziner, Doron, Thomas Stevens, Martin Stevens, Brian A. Stewart, Rebecca J. Hannagan, and Giulia Saltini-Semerari. *Biology and Philosophy.* Published online, March 10, 2010. http://www.springerlink.com.

Singer, Peter. 1981. *The Expanding Circle: Ethics and Sociobiology.* New York: Farrar, Straus and Giroux.

Smith, Adam. 1964/1776. *The Wealth of Nations.* 2 vols. London: Dent.

____. 1976/1759. *The Theory of Moral Sentiments.* Oxford: Clarendon Press.

Sober, Elliot, and David Sloan Wilson. 1998. *Unto Others: The Evolution and Psychology of Unselfish Behavior.* Cambridge, MA: Harvard University Press.

Solomon, Robert C., and Mark C. Murphy, eds. 2000. *What Is Justice? Classic and Contemporary Readings.* 2nd ed. New York: Oxford University Press.

Spector, Paul E. 2008. *Industrial Organizational Behavior.* 5th ed. Hoboken, NJ: John Wiley.

Steward, Julian H. 1938. *Basin-Plateau Aboriginal Sociopolitical Groups.* Washington, DC: U.S. Government Printing Office.

____. 1941. *Nevada Shoshone.* Berkeley, CA: University of California Press.

Stiglitz, Joseph E. 2002. *Globalization and Its Discontents.* New York: W. W. Norton.

____. 2010. *Freefall: America, Markets, and the Sinking of the World Economy.* New York: W. W. Norton.

Strum, Shirley C. 1987. *Almost Human: A Journey into the World of Baboons.* New York: Random House.

Szasz, Thomas. 1961. *The Myth of Mental Illness.* New York: Harper and Row.

Target Training International. 1990. *Personal Interests, Attitudes and Values Assessment.* Princeton, NJ: Princeton University Press.

Taylor, Charles. 1992. *The Ethics of Authenticity.* Cambridge, MA: Harvard University Press.

Taylor, John B. 2001. *Economics.* 3rd ed. Boston: Houghton Mifflin.

Thivet, Delphine. 2008. Thomas Hobbes: A Philosopher of War or Peace? *British Journal for the History of Philosophy* 16 (4): 701–21.

Thucydides. 2004/432 BC. *History of the Peloponnesian War.* Mineola, NY: Dover.

Trigger, Bruce G. 2003. *Understanding Early Civilizations: A Comparative Study.* Cambridge: Cambridge University Press.

Trivers, Robert L. 1971. The Evolution of Reciprocal Altruism. *Quarterly Review of Biology* 46:35–57.

____. 1985. *Social Evolution.* Menlo Park, CA: Benjamin/Cummings.

Turnbull, Colin. 1961. *The Forest People.* New York: Simon and Schuster.

____. 1972. *The Mountain People*. New York: Simon and Schuster.

Van Hooff, Jan A. R. A. M. 2001. Conflict, Reconciliation and Negotiation in Non-human Primates: The Value of Long-Term Relationships. In *Economics in Nature: Social Dilemmas, Mate Choices and Biological Markets*, ed. Ronald Noe, Jan A. R. A. M. van Hooff, and Peter Hammerstein, 67–90. Cambridge: Cambridge University Press.

Walster, Elaine, G. William Walster, and Helen Berscheid. 1978. *Equity Theory and Research*. Boston: Allyn and Bacon.

Walzer, Michael. 1983. *Spheres of Justice*. New York: Basic Books.

Waterfield, Robin. 2004. *Athens: A History, from Ancient Ideal to Modern City*. New York: Basic Books.

Wattles, Jeffrey. 1996. *The Golden Rule*. New York: Oxford University Press.

Webb, Sidney, and Beatrice Webb. 1920. *Industrial Democracy*. London: Longmans, Green.

Westermarck, Edward A. 1971/1906. *The Origin and Development of the Moral Ideas*. Freeport, NY: Books for Libraries Press.

Widerquist, Karl. 2005. *The Ethics and Economics of the Basic Income Guarantee*. Farnham, Surrey, UK: Ashgate.

Wilkinson, Gerald S. 1984. Reciprocal Food Sharing in the Vampire Bat. *Nature* 308:181–84.

____. 1990. Food Sharing in Vampire Bats. *Scientific American* 262 (2): 76–82.

Wilkinson, Richard, and Kate Pickett. 2009. *The Spirit Level: Why More Equal Societies Almost Always Do Better*. London: Allen Lane.

Williams, George C. 1966. *Adaptation and Natural Selection: A Critique of Some Current Evolutionary Thought*. Princeton, NJ: Princeton University Press.

____. 1992. *Natural Selection: Domains, Levels, and Challenges*. New York: Oxford University Press.

____. 1993. Mother Nature Is a Wicked Old Witch. In *Evolutionary Ethics*, ed. Matthew H. Nitecki and Doris V. Nitecki, 217–23. Albany: State University of New York Press.

Wilson, James Q. 1993. *The Moral Sense*. New York: Free Press.

Wolff, Edward N. 2000. Recent Trends in Wealth Ownership, 1983–1998. Jerome Levy Economics Institute Working Paper. 300. New York University, Department of Economics.

Wolff, Jonathan. 1996. *An Introduction to Political Philosophy*. Oxford: Oxford University Press.

Wolpoff, Milford H. 1999. *Paleoanthropology*. 2nd ed. New York: McGraw-Hill.

Woodburn, James. 1982. Egalitarian Societies. *Man*, n.s., 17 (3): 431–51.

Woodward, James, and John Allman. 2007. Moral Intuition: Its Neural Substrates and Normative Significance. *Journal of Physiology Paris* 101 (4–6): 179–202.

Young, H. Peyton. 1994. *Equity: In Theory and Practice*. Princeton, NJ: Princeton University Press.